Management by Measurement

Fiorenzo Franceschini
Maurizio Galetto · Domenico Maisano

Management
by Measurement

Designing Key Indicators
and Performance Measurement Systems

With 87 Figures and 62 Tables

 Springer

Professor Fiorenzo Franceschini
Dr. Maurizio Galetto
Dr. Domenico Maisano
POLITECNICO di TORINO
Dip. Sistemi di Produzione ed Economia dell' Azienda
Corso Duca degli Abruzzi, 24
CAP 10129
Torino
Italy

fiorenzo.franceschini@polito.it
maurizio.galetto@polito.it
domenico.maisano@polito.it

Library of Congress Control Number: 2007931051

ISBN 978-3-540-73211-2   Springer Berlin Heidelberg New York

Springer is a part of Springer Science+Business Media

springer.com

Production: LE-TEX Jelonek, Schmidt & Vöckler GbR, Leipzig
Cover-design: WMX Design GmbH, Heidelberg

SPIN 12080427   134/3180YL - 5 4 3 2 1 0   Printed on acid-free paper

Foreword

For a long time I was troubled by the doubt that those who were criticizing the School of Management Engineering were right, in considering this school a poor imitation, combining both the "old school" of engineers and that of Economics.

In spite of the success of this new professional figure into the working world, I had the irritating suspect that – in a society where everything blooms and withers rapidly – would be the result of a temporary trend, doomed to be substituted by new ones.

This perplexity (it has been difficult for me to make it clear) arose from the lack of a well-grounded and distinctive culture at the basis of Management Engineering, like the culture of the great polytechnic school of Monge and d'Alembert, never untied to the confrontation with the ability of solving new problems.

I was ignoring the fact that, letting things take their course and allowing teachers and researchers do their work, would have finally captured a new and precise identity. An identity derived from the comparison with the different, dynamic and more complex problems proposed by the actual socio-economic system, which requires – as well as the technical-scientific knowledge of the classical engineering – a more agile and flexible attitude and *modus-operandi*.

The manuscript of Franceschini, Galetto and Maisano is a concrete sign of this achievement.

The authors, by means of their robust experience in the metrological area and their long and fruitful work in the quality area, discuss the performance indicators issue, analysing it completely and organically.

Topics under discussion cross the boundary of classical engineering and experimental domain, presenting new questions and giving well-structured answers to the issues which inevitably originate from the use of indicators to evaluate results and performances in complex fields. For example within the public sector, the subject who invests and incurs expenses is not the one who evaluates and benefits from the results.

Fine are the arguments which show that indicators are not mere technical means of evaluating performance, but rather "normative" tools conditioning the behaviours of the subjects whose actions are being examined.

This mechanism – well known by sociologists, but unfamiliar to engineers – becomes an integral and integrated part of the Management Engineering culture.

Authors – real management engineers – develop the issue, not only explaining it by the use of well-fitting examples, but also suggesting the rules for the construction of performance measurement systems, identifying their potential as well as their drawbacks.

Such a text had been missing, and its appearance has made this need more clear.

In conclusion, it only remains for me to wish the authors the well-deserved success for this book.

Sergio Rossetto[1]

---

[1] Dean of the Fourth School of Engineering of Politecnico di Torino (Italy).

Preface

Every day life is literally pervaded by the presence and use of indicators: company performance indicators, price indicators, Stock Exchange indexes, air quality indicators, indicators of social status, and many others. Indicators give the impression to be the real engine of social systems, economy, and organizations. Furthermore, the interest in developing effective systems to measure performance results is growing more and more.

Is it so necessary using indicators in characterizing or evaluating complex systems/processes?

In global competition-oriented frameworks, continuous performance monitoring is not a choice. It is a need. Strategic targets and operational methods to reach and control results are necessary but, unfortunately, not sufficient conditions to ensure the survival of organizations.

In some sense, performance measurements are the core of process management. They start from collecting and analysing data, making it possible to track progress, identify strong and weak points, and − finally − drive improvements.

The purpose of this monograph is to describe in detail the main characteristics of indicators and performance measurement systems.

This text is divided into six chapters.

Chapter one deals with basic concepts about indicators and process performances. The second chapter discusses critical aspects, troubles and curiosities which can be produced representing a generic system by means of indicators. The third chapter focuses the attention on the problem of the "uniqueness" of representation. Given a process, the way to represent it through indicators is univocal? Chapter four analyses indicators properties. Description is supported by a large use of examples and practical applications. Fifth chapter illustrates methods for implementing performance measurement systems: how to activate and maintain them over time. It also examines the role of indicators as "conceptual technologies". In conclusion, chapter six deals with the concepts of indicator, measurement, preference and evaluation, comparing them from the objectivity and empiricity viewpoints.

The realization of this work is thanks to Professors Sergio Rossetto, Raffaello Levi, Quirico Semeraro, Angappa Gunasekaran, Bianca Colosimo, Grazia Vicario, colleagues and friends. We are also greatly indebted to Doctors Luisa Albano, Paolo Cecconi, Luca Mastrogiacomo, Francesca Nardilli, Elisa Turina, Gabriella Ukovich, Luciano Viticchié, for their assistance and technical support.

Turin, July 2007.

Fiorenzo Franceschini
Maurizio Galetto
Domenico Maisano

Contents

1. Quality and process indicators

1.1 General concepts

It is widely known that most of the complex organizations implement performance measurement systems, in order to give true attention to results, responsibilities, and targets.

A question arises: are indicators the "key tool" of an enterprise for optimizing process management? Organizations utilize performance indicators for many important purposes. For example, in manufacturing, sales and customer satisfaction performances make possible feeling the pulse of the market or planning the organization's future development.

Managers utilize indicators to allocate assets or to establish which strategy to implement. While *Quality* standards have become the organizations' interior operative tool, performance indicators are the *communication protocol* of their health state to the outside world. An extensive empirical research, carried out in the United States, shows that the companies winner of Quality awards are usually those with the highest profits (Singhal and Hendricks 1997).

But, how can we recognize the organizations' Quality? Quality, in its final analysis, is the ability to fulfil different types of requirements – productive, economical, social – with concrete and measurable actions. The Quality of performances is a basic element to differentiate an organization within the market.

Firstly, to make Quality concrete, we should identify the stakeholders' needs. Then it is necessary to fulfil these needs effectively, using all the essentials (processes and resources). That requires the ability to observe the evolution of the process and its context. Performance indicators are the proper tools to achieve this purpose. They are not simple observation tools. They can have a deep "normative" effect, which can modify organization behaviour and influence decisions.

If a production line manager is trained to classify as "good" those products that are spread onto the market, his attention will be directed towards maximizing the products diffusion and expansion. Unintentionally, this strategy could sacrifice long-term profits, or company investments in other products. If a *Call Center* administrator is recompensed depending on his ability in reducing absenteeism, he will try to make the absenteeism indicator decreasing, even if that will not necessarily lead to increase productivity.

The mechanism is easy to work out. If a firm measures indicators "a", "b" and "c", neglecting "x", "y" and "z", then managers will pay more attention to the first ones. Soon those managers who do well on indicators "a", "b" and "c" are promoted or are given more responsibilities. Increased pay and bonuses follow. Recognizing these rewards, managers start asking their employees to make decisions and take actions that improve these indicators and so on. The firm gains core strengths in producing "a", "b" and "c". Firms become what they measure! (Hauser and Katz 1998).

If maximizing "a", "b" and "c" leads to long-term profit, the indicators are effective. If "a", "b" and "c" lead to counterproductive decisions and actions, then indicators have failed. But even worse! Once the enterprise is committed to these indicators, indicators provide tremendous inertia. Those who know how to maximize "a", "b" and "c" fear to change the course. It is extremely hard to refocus the enterprise on new goals.

Selection of good indicators is not an easy process, with many error possibilities. This book focuses on the construction of performance measurement systems, knowing that "magic rules" to identify them do not exist. Many indicators seem right and are easy to measure, but have subtle, counter-productive consequences. Other indicators are more difficult to measure, but focus the enterprise on those decisions and actions that are critical to success. We try to suggest how to identify indicators that achieve balance in these effects and enhance long-term profitability.

The construction of a Quality System needs to consider these aspects. The first step consists in identifying stakeholders exigencies. Then, it is necessary to define performance levels, to organize and control all the activities involved in meeting the targets (practices, tasks, functions), to select indicators, to define how to gather information, and – finally – to decide on how to take corrective or ameliorative actions.

1.2 Quality Management Systems

A Quality Management System is a set of tools for driving and controlling an organization, considering all different Quality aspects (ISO-9000 2000):
* human resources;

- know-how and technology;
- working practices, methodologies and procedures.

A Quality System – with its resources and processes – should accomplish specific planned targets such as production, cost, time, return of investment, stakeholders exigencies or expectations. It can be useful for the following operations:

- performances evaluation of the whole firm aspects (processes, suppliers, employees, *Customer Satisfaction...*);
- market analysis (shares, development opportunities);
- productivity and competitors analysis;
- decisions about product innovation or new services provided.

For achieving positive results on many fronts (market shares, productivity, profit, competitiveness, customer portfolio, etc..), for each organization it is essential to implement quality management principles and methods.

The creation of Quality Management Systems is supported by eight fundamental principles in the ISO 9000:2000 Standard (ISO-9000 2000):

- *Customer Oriented Organizations.* Organizations must understand the customer needs, requirements, and expectations.
- *Leadership.* Leaders must establish a unity of purpose and set the direction the organization should take. Furthermore, they must create an environment that encourages people to achieve the organization's objectives.
- *Employees Participation.* Organizations must encourage the involvement of people at all levels, to help them to develop and use their abilities.
- *Process Approach.* Organizations are more efficient when they use a process approach to manage activities and related resources.
- *Systems Approach.* Organizations are more efficient and effective when they use a systems approach. Interrelated processes must be identified and treated as a system.
- *Continuous Improvement.* Organizations must make a permanent commitment to continually improve their overall performance.
- *Facts before decisions.* Organizations must base decisions on the analysis of factual information and data.
- *Partnership with Suppliers.* Organizations must maintain a mutually beneficial relationship with their suppliers to help them create value.

These principles should be applied to improve organizational performance and achieve success. The main benefits are the following:

- marketing and customer relations benefits:
  - support for new products development;
  - easier access to market;
  - customers are aware of organizations efforts into research and quality;
  - better credibility of organizations.
- internal benefits:
  - Quality is easier to plan and control;
  - support for the definition of internal standards, working practices, and procedures;
  - more effective and efficient working operations.
- benefits for the relationships with suppliers:
  - better integration with suppliers;
  - reduction of the number of suppliers and use of rational methods for their evaluation and selection;
  - support to search for new suppliers.

1.3   The concept of process

1.3.1 Definition

According to the ISO 9000:2000 standard, a process is *"an integrated system of activities that uses resources to transform inputs into outputs"*. This general definition identifies the process like a black box, in which input elements are transformed into output(s).

The process approach is a strong management tool. A system is generally made of several processes interconnected: the output from one process becomes the input for other ones. Processes are "glued" together by means of such input-output relationships. When analysing each process, it is necessary to identify the output target characteristics, and who will benefit from them. Not only final users, but all the functions involved in the processes – inside and outside the firm – should be considered.

Increasing the analysis level of detail, each process can be broken-down into sub-processes, and so on. This sort of "explosion" should be continued, in order to identify all the organization basic components.

Monitoring a process requires identifying specific activities, responsibilities, and indicators for testing effectiveness and efficiency. *Effective-*

ness means setting the right goals and objectives, making sure they are properly accomplished (*doing the right things*). Effectiveness is measured comparing the achieved results with target objectives. On the other hand, *efficiency* means getting the most (output) from your resources (input), whether they are people or products (*doing things right*). Efficiency defines a link between process performances and the resources employed.

1.3.2 Process modeling

Process managing needs a proper modelization, which considers major activities, decision-making practices, interactions, constraints, and resources. It is important to decide which process characteristics to emphasize and then represent.

Modeling a process means describing it, considering the targets which should be met. Process is a symbolic "place" where consumer expectations are turned into firm targets, and targets into operative responses. A proper performance measurement system should be arranged to verify if responses are consistent with requirements.

Modelization techniques/methodologies should be able to highlight process characteristics and peculiarities (organizational, technological, relational aspects…). These methodologies are generally supported by software applications which map and display activities/actors involved and focus the attention on many process aspects (input, output, responsibilities, etc..) and practical parameters (time, cost, constraints, etc..).

Mapping is essential to "understand" the process. Furthermore, it is possible to perform process performance simulations, identifying "optimal" operation conditions in terms of costs, time, and quality. A significant support to managers is given by the processes representation tools. IDEF, CIMOSA, DSM, etc… are some of the most used (CIMOSA 1993; Draft Federal Information 1993; Mayer et al. 1995; Ulrich and Eppinger 2000).

These methodologies make it possible to merge different accessory "views" of the organization: functions, activities, resources, and physical/informative flows.

1.3.3 Process "measurement"

The object of process construction is to meet stakeholder needs. Consequently, it is essential to set up a measurement system to test this condition. Identifying and controlling process performance and evolution are indispensable actions taken to decide which strategies carry out.

According to the UNI 11097:2003 Standard (UNI-11097 2003):

"A system of indicators should become an information system for esti- mating the level of achievement of quality targets".
Indicators selection should be performed considering:

- quality policy;
- quality targets;
- the area of interest, within the organization: market competitiveness; customer satisfaction; market share; economical/financial results; quali- ty, reliability, and service; flexibility of factory systems and services supply; research and development; progress and innovation; manage- ment, development and enhancement of human resources; internal and external communication);
- performance factors;
- process targets.

It must be emphasized that any deficiency of the process measurement system will affect the so-called non-quality costs. These costs are the most powerful and rational lever to persuade organizations to apply themselves on continuous improvement.

Process implementation should be followed by a systematic monitoring plan and periodical performance recording, in order to identify process critical aspects and to reengineer process activities. Fig. 1.1 represents this concept.

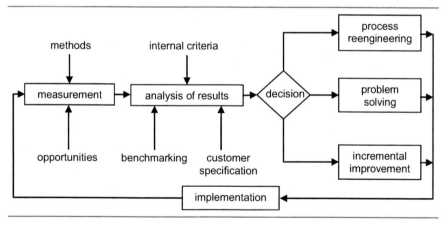

Fig. 1.1. The process improvement chain (Barbarino 2001). With permission

The main activities that process measurement systems entail are:

- *Definition of indicators.* This phase consists in defining which perfor- mance measurements should be used and which data should be collected. Measured parameters are selected depending on the process critical aspects and growth potential.

- *Decision*. Depending on the difference between target and measured performance level, there are three different courses of action:
  - individual problem solving;
  - incremental improvements (*step by step*);
  - process reengineering.

As shown in Fig. 1.1, the performance measurement system is directly connected to the "self-regulating" process chain. Process outputs are turned into input for the performance measurement system, which is a spur for possible actions or decisions. The main difficulties of this approach consist in implementing the performance measurement system and providing its continuity.

The firm management is the final receiver of monitoring processes. Analysis results are used to perform detailed evaluations and to take decisions concerning the attribution of responsibility, or the assignment of resources. These decisions may influence the firm's future behaviour.

Measurements should be technically and economically efficient, and should focus the attention on results instead of actions: the purpose of doing a task takes the precedence over the way with which to carry it out. For that reason, most of the monitored variables should be easy to measure: quantitative variables are more practical than qualitative ones. The firsts can be easily referred to monetary values, instead of the seconds – which are more efficient to detect consequences of actions and/or firm behaviours – but more difficult to be assessed in terms of money (Bellandi 1996).

1.4   Process indicators

As explained before, measuring is essential for the process performance control and improvement. However, constructing and starting-up a measurement system is easier said than done. The most critical aspect does not consist in identifying the indicators, but rather in identifying those that "properly" represent the process: the so-called *Key Performance Indicators (KPI)*.

The UNI 11097 Standard classifies as quality indicator *"the qualitative and/or quantitative information on an examined phenomenon (or a process, or a result), which makes it possible to analyze its evolution and to check whether quality targets are met, driving actions and decisions"* (UNI-1097 2003).

This definition identifies several critical elements: (1) indicators definition; (2) indicators should be well-understood and accepted by process managers and employees; (3) indicators traceability and verifiability.

Generally, each indicator refers to a specific target, that is to say a sort of reference point used as a basis of comparison. This reference can be *absolute* or *relative* (depending on whether it is external or internal to the organization). A "zero defects" program is an example of absolute reference. Reference values can be derived from the organization's past experience or – when it is not feasible – they can be extrapolated from similar processes (*benchmarking*).

The definition of indicator, given by UNI 11097, entails some basic requirements:

• representativeness;
• simple and easy to interpret;
• capable to indicate time-trends;
• sensitive to changes within or outside the organization;
• easy data collecting and processing;
• easy and quick to update.

Relevant characteristics and properties of indicators are discussed on Chap. 4.

The most significant aspects for characterizing the state of a process are defined *quality factors* (or *analysis dimensions*). Each of them should be identified and associated to one or more process indicator(s).

The UNI 11097 Standard continues, explaining that "*measurements of the examined phenomenon should be faithful and properly documented, without any distortion or manipulation. The information provided by the indicator should be exact, precise and sensitive to significant changes, as well as steady to be reproducible. Furthermore, with the aim of simplifying the analysis and the synthesis, information should be quantified during data collection.*"

One of the most difficult activities in process management is making systems "tangible", by means of their performances. Process managers try to do this, translating organization goals into different metrics (indicators), which are also visible from the outside. This *modus operandi* is usual for each type of system: a manufacturing process, a service, or a generic organization. The typical question asked by the process managers is: "*Do process performances meet the expected targets?*" (Magretta and Stone 2002).

Choosing the right indicators is a critical aspect in translating an organization's mission, or strategy into reality. Indicators and strategies are tightly and inevitably linked to each other. A strategy without indicators is useless; indicators without a strategy are meaningless.

The interest towards indicators is increasing. Their importance has been long recognized in all contexts, but their faults a little less. This manuscript tries to highlight the potential of indicators, as well as their major drawbacks.

Every firm, every activity, every worker needs indicators. Indicators fulfil the fundamental activities of measuring (evaluating how we are doing), educating (since what we measure is what is important; what we measure indicates how we intend to deliver value to our costumers), and directing (potential problems are flagged by the size of the gaps between the indicators and the target).

Yet, performance measurements continue to present a challenge to operations managers as well as researchers of Operations Management. Operating indicators are often poorly understood and guidelines for the use of indicators are also poorly articulated.

While there are numerous examples of the use of various indicators (indicators are currently used in several area of interest – like Logistics, Quality, Information Sciences, System Engineering), there are relatively few studies in Operations Management that have focused on the development, implementation, management, use and effects of indicators within either the operations management system or the supply chain. Nascent examples can be found in the research of Beaumon (1999), Leong and Ward (1995), Neely (1998), and New and Szwejczewski (1995). A great deal of what we currently know about indicators comes from the managerial literature, e.g. (Brown 1996; Dixon et al. 1990; Kaydos 1999; Ling and Goddard 1988; Lockamy III and Spencer 1998; Lynch and Cross 1995; Maskell 1991; Melnyk and Christensen 2000; Smith 2000).

We should point out that the topic of indicators as discussed by managers differs from the topic of measurement typically discussed by academics. This is primarily a by-product of different priorities between these groups. The academic is concerned with defining, adapting and validating measures to address specific research questions. The time required to develop and collect the measures is of less importance than the validity and generalizability of the results beyond the original context. Managers face far greater time pressures, and are less concerned about generalizability. They are generally more than willing to use a "good enough" measure, if it can provide useful information quickly. However, as long as the difference in priorities is recognized, the two points of view are gradually becoming closer. Undoubtedly academic measurement experts can contribute to managers' understanding of indicators, as well as managers' practical exigencies, which can be useful for academics in studying the usefulness of indicators and measuring procedures.

Recent studies suggest that indicators and performance measurement, viewed as a strategic tool for process management, are receiving more attention. Many research programs, all over the world, are dealing with these questions. For example, KMPG, an international private company, in conjunction with the University of Illinois undertook a major research initiative aimed at funding research in performance measurement (to the tune of about US$ 3 million).

The January 2003 *Harvard Business Review* case study focused on the miscues and disincentives created by poorly thought out performance measurement systems (Kerr 2003). Why is there an increasing interest? Here are some possible reasons:

- "never satisfied" consumers (McKenna 1997);
- the need to manage the "total" supply chain, rather than only internal factors (holistic vision);
- shrinking of products/services life cycle;
- more (but not necessarily better) data;
- an increasing number of decision support tools which utilize indicators.

These dynamics make "static" indicators system obsolete and call for new performance measures and approaches, which need to go beyond simple reporting, by identifying improvement opportunities and anticipating potential problems. Furthermore, indicators are now viewed as an important means, by which priorities are communicated within the firm and across the supply chain. Indicators misalignment is thought to be a primary source of inefficiency and disruption in supply chain interaction.

1.4.1 Indicators functions

Indicators provide a means of "distilling" the larger volume of data collected by organizations. As the volume of inputs increases, through greater span of control or growing complexity of an operation, data management becomes increasingly difficult. Actions and decisions are greatly influenced by indicators nature, use, and time horizon (short or long-term).

Indicators provide the following three basic functions:

- *Control*. Indicators enable managers and workers to evaluate and control the performance of the resources which they are responsible.
- *Communication*. Indicators communicate performance not only to internal workers and managers for a purposes of control, but also to external

stakeholders for other purposes. Poorly developed or implemented indicators can lead to users feeling frustrated and confused.

- *Improvement.* Indicators identify gaps (between performance and expectation) that ideally point the way for intervention and improvement. The size of the gap and its direction (positive or negative) provide information and feedback, which can be used to identify productive processes adjustments or other actions.

Each system of indicators is subject to a dynamic tension. This stems from the desire to change indicators in response to new strategic priorities, and the desire to maintain indicators to allow comparison of performance over time. This tension will dictate the indicators *life cycle*.

1.4.2 Aims and use of indicators

Regarding the study of indicators, one source of complexity is the variety of different types of indicators that researchers and managers encounter. Various indicators can be classified according to two primary attributes: *indicator focus* and *indicator tense*.

Indicator focus pertains to the resource that is the focus of the indicator. Generally, indicators report data in either financial (monetary) or operational (e.g., operational details such as lead times, inventory levels or setup times) terms. Financial indicators define the pertinent elements in terms of monetary resource equivalents, whereas operational indicators tend to define elements in terms of other resources (e.g., time, people) or outputs (e.g., physical units, defects).

The second attribute, indicator tense, refers to how the indicators are intended to be used. Indicators can be used both to judge outcome performance (*ex post*) and to predict future performance (*ex ante*). Many of the cost-based indicators encountered in firms belong to the first category. In contrast, a predictive use of an indicator is aimed at increasing the chances of achieving a certain objective or goal. If our interest is in reducing *lead time*, then we might assess indicators such as distance covered by the process, *setup* times, and number of steps in the process. Reductions in one or more of these areas should be reflected in reductions in *lead time*.

An emphasis on identifying and using indicators in a predictive way is relatively new. Predictive indicators are appropriate when the interest is in preventing the occurrence of problems, rather than correcting them.

The combination of these attributes (focus and tense) provides four distinct types of indicators, as shown in Fig. 1.2.

Each cell in Fig. 1.2 scheme identifies a particular use for each indicator. These different categories appeal to different groups within the firm.

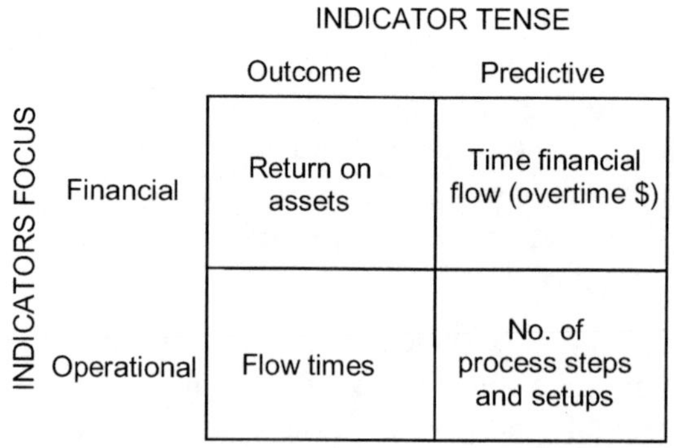

INDICATOR TENSE

	Outcome	Predictive

		Outcome	Predictive
INDICATORS FOCUS	Financial	Return on assets	Time financial flow (overtime $)
	Operational	Flow times	No. of process steps and setups

Fig. 1.2. Classification of indicators on the basis of focus and tense attributes (Melnyk et al. 2004). With permission

Top managers, for example, are most typically interested in financial/outcome. In contrast, operations managers and workers are most likely to be interested in operational/predictive or operational/outcome indicators. Since these are two sets of operation which pertain to the processes that the managers must manage and change.

1.4.3 Terminology

The terminology used in the performance measurements context is not completely and univocally defined. Often, similar concepts are classified using different terms, depending on the technical area of interest. For example, terms such as "metric" and "performance indicator" are usually considered as synonyms. The same happens for terms such as "target", "result" or "performance reference". In the following descriptions we try to clarify the meaning of each of these specific terms. When possible, our terminology will refer to the ISO 9000:2000 Standard.

1.4.4 Categories of indicators

The term "indicator" is often used to refer to one of the following categories:

- the basic indicators;
- the derived indicators;
- the *indicators sets*;
- the overall performance measurement systems.

These types of indicators are linked to each other. At the base there are the *basic indicators*, the "building blocks". Basic indicators are aggregated in order to form: *indicators sets* and *derived indicators* – which are the synthesis of two or more indicators.

Each *set of indicators* represents and regulates a specific process function. Global process management and coordination are carried out by the *performance measurement system*, which is at the highest level of the hierarchy. Performance measurement system is responsible for coordinating indicators across the various functions, and for aligning the indicators from the strategic (top management) to the operational (shop floor/purchasing/execution context) levels.

For every activity/product/function, multiple indicators can be developed and implemented. The challenge is to design a structure to the indicators (i.e. grouping them together) and extracting an overall sense of performance from them.

In the current literature, several different approaches have been proposed for developing such an integrative system. These include (see Chap. 5):

- the *Balanced Scorecard* method, examined in Chap. 5 (Kaplan and Norton 1992, 1996, 2001; Ittner and Larcker 1998);
- the *Strategic Profit Impact* model, also known as the *Dupont* model (Lambert and Burduroglu 2000);
- the *Critical Few* method (Performance-Based Management Special Interest Group 2001);
- the models *EFQM (European Foundation for Quality Management)*, and *Malcom Baldrige Quality Award* (EFQM 2006; BNQP 2006; NIST 2006).

Each of these major systems has strengths and weaknesses. For example, the *Balanced Scorecard* excels as it is able to force Top Management to recognize that multiple activities must be carried out for the firm to succeed. The management and monitoring of these activities must be balanced. All the firm's features (dimensions) should be considered, not only the economical ones. Furthermore, this model gives useful information on how to perform a practical indicators' synthesis, useful for organizational activities.

The performance measurement system is ultimately responsible for maintaining *alignment* and *coordination*. Alignment deals with the maintenance of consistency between the strategic goals and indicators. Alignment attempts to ensure that the objectives set at the higher levels are consistent with and supported by the indicators and activities of the lower levels. In contrast, coordination recognizes the presence of interdependency between processes, activities or functions. Coordination deals with the degree to which the indicators in various related areas are consistent with each other and are supportive of each other. Coordination strives to reduce potential conflict when there are contrasting goals. For example, in manufacturing, productivity indicators (number of elements produced) conflict with quality indicators (number of defects).

A good indicator set directs and regulates the activities in support of strategic objectives and provides real-time feedback, predictive data, and insights into opportunities for improvement. In addition, indicators need to be flexible in recognising and responding to changing demands placed on the operating system due to product churn, heterogeneous customer requirements, as well as changes in operating inputs, resources, and performance over time.

The coming chapters will deepen these themes, paying particular attention to indicators potentials and limitations.

1.4.5 A general classification of indicators

Indicators should provide accurate information about the status and the possible changes of a process. UNI 11097 Standard (see Sect. 3.3) suggests an interesting classification, depending on the process "observation moment".

There are three types of indicators, which are individually discussed in the three following sections:

- *Initial* indicators. Indicators of the quality of materials or the quality of services provided by suppliers.
- *Intermediate* indicators. For example, indicators of a manufacturing process compliance.
- *Final* (result) indicators. For example, indicators of customer satisfaction or production cost.

Initial indicators (or structure indicators)

Planning is the first task in a project. It makes it possible to estimate if the organization is able to meet its targets, considering the available resources (organizational, physical and monetary).

Initial indicators – or structure indicators – give an answer to the question *"what are the process available assets and the working patterns?"*, considering all the resources involved: facilities, human resources, technological and monetary assets, services provided by suppliers, and so on.

These indicators are also used to qualify the skill and involvement level of the personnel. The final purpose is to provide a clear indication in order to improve the project planning/management.

Intermediate indicators (or process indicators)

Intermediate indicators give an answer to the question *"how the process works?"* They measure the consistency between process results and process specifications, providing useful information on the process state. This type of control makes it possible to understand whether process conditions are stable or, on the contrary, whether process has run into unexpected or unpredictable difficulties.

Final indicators or (result indicators)

Final indicators – or result indicators – answer to the following questions:

- "What are process outcomes?"
- "Has the process met the purposes?"
- "What are expected/unexpected effects produced by the process?"
- "What is the cost-benefit ratio?"

Final indicators are generally viewed as the most important ones, since they estimate process final results, both positive and negative. For example, they may deal with customer satisfaction or with the production cost of products/services.

A second indicators classification is based on their "position" within the organizational framework. Fig. 1.3 represents the pyramidal categorization suggested by Juran (Juran 2005).

At the pyramid bottom, there are "technological measurement systems", for the monitoring of the parts of products, processes and services. At second level, indicators "synthesise" basic data on individual product or process: for instance the percentage of defects in a specific product part or service.

Third level includes "quality measurement systems", dealing with entire sectors, production lines or services. At the top of the pyramid, we find the "overall synthesis indicators", used by top management to evaluate the whole conditions of economic/monetary aspects, manufacturing processes, and market.

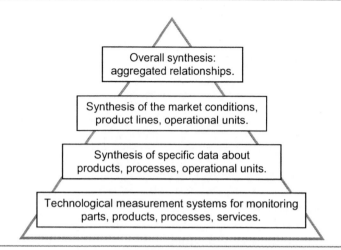

Fig. 1.3. The measurement systems pyramid (Juran 2005; Bellandi 1996). With permission

1.4.6 Comparison between economic and process indicators

Economic-financial indicators

Process performances of each organization can hardly be monitored without using "monetary" indicators.

Indicators derived from the general accounting are traditionally used to measure performances. These indicators are generally easily measurable and user-friendly.

Since economic outcomes are the result of past decisions, economic indicators can not be used to identify future opportunities of an organization. Classically, the most common drawbacks of economical-financial performance indicators are:

- they are not prompt; the evaluation and aggregation of physical transactions may require a lot of time (especially for firms with a large range of products);
- they preferentially report information to the outside, rather than to the inside of the firm;
- they focus on costs, in particular direct labour costs, which are nowadays less decisive in the determination of the processes added-value;
- they ignore quality, innovation potential, competencies, skills improvement, and the strategic dimensions of competitiveness and added-value;

- sometimes they slow down the development of new and more suitable organizational structures;
- they are not very sensitive to changes in the firm's strategies, or in the outside system.

Reduced timeliness is the major limitation of financial indicators. To determine them, all the information (market shares, product and process characteristics, etc..) needs to be translated into monetary terms. While economic indicators can be calculated from the final balance, financial indicators need to estimate future results. Consequently, they require a more complex and extended analysis. This reduces the possibility of performing frequent data collecting to identify problems promptly.

One of financial indicators strengths is long-term orientation, which derives from the joint analysis of short-term and long-term results.

Considering financial indicators, the link between indicators completeness and precision depends on the analysis level. The more strategic aspects are examined in detail, the more the analysis will result complete.

The use of general accounting indicators should be limited to firms operating within a stable context, where short-time profitability may properly represent their competitiveness. When the context is dynamic, it is more difficult to identify a correlation between past and future results. For that reason, the more the context is dynamic, the more crucial it becomes for the information to be timeliness (Azzone 1994).

There are two possible solutions to overcome these limitations: either by improving the current financial indicators, or by focusing attention on the operational measurements. As a matter of fact, managers' analysis is based on more than one indicator, not to run the risk of forgetting some critical aspects for the business. A possible solution consists in constructing a balanced representation of both financial and operational measurements (Kaplan and Norton 1992, 1996).

Process indicators

Process indicators can be classified depending on the measured *competitiveness factor* (time, quality, flexibility, productivity, and environmental compatibility), and their purpose (indicators of customer satisfaction, or indicators for the internal resources managing).

Time measurements typically relate to the process of time development, for example the time for product development or the time for logistical development process.

Time indicators can be divided into two main categories. The first sees time as a source of internal efficiency; in this case, time saving means cost reduction and creation of value. The second category includes those indicators related to the timeliness in market response. In this case, time is viewed as a lever for the product differentiation and the increase of proceeds.

Internal (time) indicators aim at identifying process activities which produce added-value, such as those for improving product performance perceived by costumers. On the other hand, external (time) indicators can be divided into:

- standard products delivery timeliness (indicators aimed to evaluate the logistical system competitiveness);
- new products development time (indicators aimed to evaluate the competitiveness of the products development process). The most common indicator is *time-to-market* (time period between product concept and product launch onto the market).

Quality measurements investigate the product/service characteristics compared to customer needs (accordance with product specifications, correspondence with customer needs) and to the process efficiency/effectiveness criteria (resources waste, defectiveness, etc...).

Productivity measurements are represented by the classical process indicators. They are defined as the ratio of process outputs to process inputs, and are primarily used to estimate the labour productivity. Productivity indicators are typically used in processing industries, where output results can be easily measured.

Measurements of environmental compatibility aim at identifying the firm's ability to develop environmentally friendly products/processes. Although in the recent past they were restricted only to the technical/operational field, environmental issues become strategically more and more important for the firm, influencing the creation of value.

Flexibility measurements evaluate the firm's ability to quickly respond to changes, keeping down time and cost. Consequently, the more the context is dynamic, the more importance is given to flexibility. There are two typologies of changes: quantitative changes – related to positive or negative fluctuations in products/services demand – and qualitative changes – related to modifications in products/services typologies. Depending on the type and size of changes, we can identify six flexibility's dimensions: volume, mix, modularity, product, production and operation.

The major distinctive element of process indicators is *timeliness*. While economic indicators entail that physical transactions are translated into monetary terms, process indicators simply derive the information from transactions.

A second distinctive element is *long-term orientation*. Process indicators may provide a synthesis of the firm's competitive advantages.

Furthermore, it is difficult to assess the *completeness* of process indicators. On the contrary, economic-financial indicators aggregate and synthesise several performances into a single monetary variable. On the other hand, each process indicator is related to a specific type of performance: a competitive time-to-market does not guarantee that the product quality will satisfy customers. A limitation of these indicators consists in losing sight of the firm's whole complexity (Azzone 1996).

1.4.7 Indicators and research: the state of the art

The "hot" front of the research on performance indicators is the study of their impact on to complex systems.

The topic is not completely new. Skinner in 1974 identified simplistic performance evaluation as being one of the major causes for factories getting into trouble (Skinner 1974). Subsequently, Hill (1999) recognized the role and impact of performance measures and performance measurements systems in his studies of manufacturing strategy. In these and other studies, indicators are often viewed as being part of the infrastructure or environment in which manufacturing must operate (*conceptual technologies*).

However, while we have recognized the role of indicators as an influencing factor, there is still a need to position the topic of indicators within a theoretical context – a framework that gives indicators a central role.

One such theoretical framework is *Agency Theory* (Eisenhardt 1989). Agency theory applies to the study of problems arising when one party, the principal, delegates work to another party, the agent. The unit of analysis is the metaphor of a contract between the agent and the principal. What makes agency theory so attractive is the recognition that in most organizations the concept of a contract as a motivating and control mechanism is not really appropriate. Rather, the contract is replaced by the indicator. It is the indicator that motivates and directs; it is the indicator that enables principals to manage and direct the activities of their various agents (Austin 1996).

Another interesting framework for future research is the *Dependency Theory* (Pfeffer and Salancik 1978). This theory states that the degree of interdependence and the nature of interactions among functional specialists within an organization are influenced by the nature of the collective task which they seek to accomplish. In dynamic environments, involving rapid product change and high degrees of heterogeneity in customer requests, agents responsible for different functional aspects of order taking, processing, and fulfilment become more and more dependent on each other for information necessary to complete their respective tasks. Dependency theory has implications for the design of indicators systems. For example it can be helpful to provide an answer to questions such as: "How should indicators reflect the interdependencies of different functional areas?", or "How should the rotation or change in indicators be associated to the dynamics of demands placed on the operating system?"

A third potentially rewarding way to look at indicators is offered by Galbraith (1973). The basic idea is that, presumably, a richer "indicators set" creates the basis for richer communications among decision makers, workers, strategy representatives, and customer of a process. However, there may be limits to the organization's (as well as individuals') ability to process larger sets of indicators. Increasing numbers of indicators could lead to greater conflict in the implied priorities, as well as greater equivocality regarding future actions. Given this apparent trade-off between indicators set richness and complexity, an information processing theoretical view could stimulate research into question regarding the optimal size of an indicators set, or perhaps the optimal combination of outcome and predictive indicators included in the set.

A further research issue, which may be added to the previous ones, concerns with the verification of the *condition of uniqueness*: "*Given a process, is there a unique set of indicators properly representing it?*" Chap. 3 will provide an answer to this central question.

In conclusion, other research topics include:

- evaluating the relationship between financial and operating indicators;
- measuring performance within the supply chain environment;
- assessing consistency of indicators, both among themselves and between indicators and the corporate strategy;
- implementing *dynamic* (over time) performance measurement systems;
- integrating performance measurement with the real or perceived reward/incentive structure.

2. Indicators criticalities and curiosities

2.1 Introduction

Indicators can be used within a wide-range, as discussed in Chap. 1. By reading any journal, it seems that these "magic" numbers influence the fate of the world: "European countries with deficit/GDP ratio lower than 3% can adopt Euro currency"; "country inflation is running at 2.7 %"; "the air quality index value is 6 (the elderly and children may be at risk. It is advised that these categories of people limit prolonged periods of time outdoors)" and so on.

Why are these indicators considered so important? Presumably, since they are supposed to represent reality.

On the concept of *representation faithfulness* we will carry out a deep analysis. Anyway, when systems to be monitored become complex, the use of indicators is practically inevitable. Let us consider, for example, the HDI (*Human Development Index*) indicator, introduced by the United Nations Development Programme (UNDP) to measure the world countries development (United Nations Development Programme 2003; Bouyssou et al. 2000).

Is the information provided by the indicator independent of the context in which it is used? In other words, is the indicator influenced by the application context or by the subjects who use it all over the world (territory planners, administrators, firm persons responsible, etc...)?

In the 1997 annual report, UNDP (1997) cautiously states that "...*the HDI has been used in many countries to rank districts or counties as a guide to identifying those most severely disadvantaged in terms of human development. Several countries, such as the Philippines, have used such analyses as a planning tool. (...) The HDI has been used especially when a researcher wants a composite measure of development. For such uses, other indicators have sometimes been added to the HDI ...*"

What has HDI been created for? What are its goals? To exclude aid from those countries which do not correctly plan development? To divide the International Monetary Fund aid among poorer countries? Are we sure that HDI is properly defined (according to its goals)? Are HDI results significant?

Trying to answer these questions, the following sections provide a detailed discussion about three distinct typologies of indicators: HDI, the air quality indicators, and the scoring indicators used for Olympic decathlon (Bouyssou et al. 2000). We will analyse each of them to identify the construction methods, and to figure out what their specific qualities and drawbacks are.

2.2 HDI indicator

The HDI is a measure to summarize human development. It measures the average achievements in a country, considering three basic dimensions of human development (UNDP 2003)[2]:

- a long and healthy life, as measured by life expectancy at birth (*Life Expectancy Index* - LEI);
- knowledge (Educational Attainment Index - EAI), as measured by the adult literacy rate (*Adult Literacy Index* - ALI), which account for 2/3, and the combined primary, secondary and tertiary gross enrolment ratio (ERI), which account for 1/3 of the total amount;
- a decent standard of living, as measured by GDPI (*Gross Domestic Product per capita Index*) given as *Purchasing Power Parity* US$ (PPP US$).

Before the HDI is calculated, an index needs to be created for each of these dimensions. To calculate these three dimension indices, minimum and maximum values are chosen for each underlying indicator. Performance in each dimension is expressed as a value between 0 and 1 by applying Eq. 2.1:

$$\text{Dimension index} = \frac{\text{actual value - minimum value}}{\text{maximum value - minimum value}} \tag{2.1}$$

Table 2.1 reports the limits for calculating the HDI.

The HDI is then calculated as a simple average of each dimension indices.

---

[2] Since 1990, *Human Development Report* (HDR) is the annual publication of the *United Nations Development Programme* (UNDP). The 2003 HDR refers to data collected in 2001.

Table 2.1. Limits for calculating the HDI

Indicator	Name	Unit of measurement	Upper limit	Lower limit
Life expectancy at birth	LEI	Years	85	25
Adult literacy rate	ALI	%	100	0
Combined gross enrolment ratio	ERI	%	100	0
GDP per capita	GDPI	PPP$	40000	100

The following sections illustrate the calculation of the HDI for Albania (data of 2001), which will be the sample country.

2.2.1 Life Expectancy Index (LEI)

The life expectancy index (LEI) measures the relative achievement of a country in life expectancy at birth. For Albania, with a life expectancy of 73.4 years in 2001, the life expectancy index is:

$$\text{LEI} = \frac{73.4 - 25.0}{85.0 - 25.0} = 0.807 \qquad (2.2)$$

2.2.2 Educational Attainment Index (EAI)

The education index (EAI) measures a country's relative achievement in both adult literacy and combined primary, secondary and tertiary gross enrolment. First, an index for adult literacy (ALI) and one for combined gross enrolment (ERI) are calculated (data are referred to the school year 2000/2001).

$$\text{ALI} = \frac{85.3 - 0.0}{100.0 - 0.0} = 0.853 \qquad (2.3)$$

$$\text{ERI} = \frac{69.0 - 0.0}{100.0 - 0.0} = 0.690 \qquad (2.4)$$

Then, these two indices are combined to create the education index, adult literacy accounting for 2/3 and the combined gross enrolment accounting for 1/3. For Albania, with an adult literacy rate of 85.3% in 2001 and a combined gross enrolment ratio of 69% in the school year 2000/01, the education index was 0.799.

$$EAI = \frac{2ALI + ERI}{3} = \frac{2 \cdot 0.853 + 0.69}{3} = 0.799 \qquad (2.5)$$

2.2.3 Gross Domestic Product Index (GDPI)

The GDP index is calculated using adjusted GDP per capita (PPP US$). Considering the HDI, income represents all the remaining dimensions of human development except long/healthy life and knowledge. Since the value of one dollar is different for people earning 100 $ in comparison to those earning 100 000 $, the income has been adjusted (*marginal utility* concept). The income adjustment function is the logarithm function (UNDP 2003). Fig. 2.1 shows the effect of the adjustment: the same increase in the adjusted income (*Log* GDP per-capita) determines a little shift of GDP per-capita when the income is low, and a high shift when the income is high.

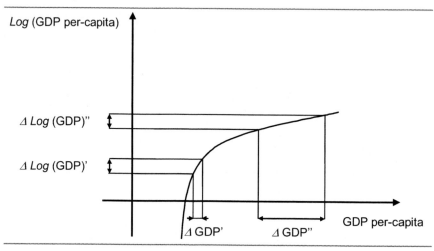

Fig. 2.1. Concept of marginal utility of the income per-capita. The same increase in the adjusted income function − *Log*(GDP per-capita) − determines a little shift of GDP per-capita when the income is low, and a high shift when the income is high

GDPI is calculated using the following formula:

$$GDPI = \frac{Log(\text{GDP pro-capite}) - Log\ 100}{Log\ 40000 - Log\ 100} \qquad (2.6)$$

For Albania, with a GDP per capita of $ 3 680 (PPP US$) in 2001, the GDP index is (UNDP 2003):

$$GDPI = \frac{Log\ 3680 - Log\ 100}{Log\ 40000 - Log\ 100} = 0.602 \tag{2.7}$$

2.2.4 Calculating the HDI

Once the dimension indices have been calculated, determining the HDI is straightforward. It is a simple average of the three dimension indices:

$$HDI = \frac{LEI + EAI + GDPI}{3} = \frac{LEI + (\frac{2ALI + ERI}{3}) + GDPI}{3} \tag{2.8}$$

For Albania, the HDI is:

$$HDI = \frac{0.807 + 0.798 + 0.602}{3} = 0.736 \tag{2.9}$$

2.2.5 Remarks on the properties of HDI

Scales normalization

To calculate HDI, the performance in each underlying indicator (LEI, ALI, ERI and GDPI) is normalized within the interval [0, 1]. Lower and upper limits, for each dimension (Table 2.1), are quite arbitrary. Why have the life expectancy limit values been set at 25 and 85 years? Is 25 years the minimum value registered? In actual fact, the lowest value ever registered is 22.6, related to Rwanda (UNDP 1997). In this case the LEI value is negative. The lower limit has been set at 25 years, at the time of the first UNDP report (1990), when the lowest value registered was 35. In that period, nobody imagined that the expectancy value could fall below the 25 years limit. To avoid this problem, the limit could have been set at a smaller value (for example 20 years).

It is interesting to notice that the choice of reference values has had a direct consequence on the HDI calculation. Let consider, for example, the conditions in Table 2.2, reporting the values of LEI, EAI and GDPI for Turkmenistan and Georgia (UNDP 2003).

If LEI minimum and maximum values are set at 25 and 85, then HDI is 0.748 for Turkmenistan, and 0.746 for Georgia. If the maximum moves to 80, HDI is respectively 0.769 for Turkmenistan and 0.770 for Georgia. This simple adjustment reverses the two countries human development indexes.

Table 2.2. Life expectancy, EAI and GDPI for Turkmenistan and Georgia (UNDP 2003)

Country	LEI	EAI	GDPI
Turkmenistan	66.6	0.92	0.63
Georgia	73.4	0.89	0.54

This example shows how tricky the definition of the same indicator reference values can be. By simply changing them, the indicator can guide to a different conclusion.

In practical terms, reducing the life expectancy interval from [25, 85] to [25, 80] makes the LEI values increase with a growth factor of (85 - 25)/(80 - 25)≈109%. As a consequence, considering HDI, the influence of LEI grows at the expenses of EAI and GDPI. Since the Georgia LEI value is greater than Turkmenistan's, their HDI positions becomes overturned.

Equally, we can state that ALI and ERI limits – respectively set at 0 and 100 – are arbitrary too. These values are not likely to be observed in a reasonably recent future. So, the *real* interval is tighter than [0, 100], and the scale could be normalized using other reference values.

The effects of compensation

Let us consider Table 2.3, reporting the values of LEI, ALI, ERI and GDP per-capita for Peru and Lebanon (UNDP 2003). Peru's indicators are greater than Lebanon's, except for the LEI. In our opinion, this is a clear sign of underdevelopment for Peru, even if the other indicators are rather good. However, the value of HDI is the same for both countries (0.752). The result is due to the effect of compensation among HDI sub-indicators (dimensions). For both of these countries, weak sub-indicators are compensated by strong ones, so that the HDI final values are the same.

Table 2.3. Values of LEI, ALI, ERI and GDP per-capita for Peru and Lebanon (UNDP 2003)

Country	LEI	ALI	ERI	GDP per-capita
Peru	69.4	0.902	0.83	4570
Lebanon	73.3	0.865	0.76	4170

The compensation effect is not reasonable when extremely weak sub-indicators are compensated by others extremely excellent. To what extent is this sort of "balancing" correct?

Let us consider LEI and GDP per-capita. A one-year decrease of life expectancy can be compensated by an increase of GDP per-capita.

A one-year life expectancy decrease corresponds to:

$$\Delta(LEI) = \frac{1}{85\text{-}25} = 0.0167 \tag{2.10}$$

It can be compensated by an increase of GDP per-capita (X), corresponding to:

$$\Delta(LEI) = \Delta(GDPI) = \Delta\left(\frac{LogX - Log100}{Log40000 - Log100}\right) = 0.0167 \tag{2.11}$$

from which we obtain:

$$\Delta\left(\frac{LogX - Log100}{Log40000 - Log100}\right) =$$
$$= \left(\frac{LogX - Log100}{Log40000 - Log100}\right) - \left(\frac{LogX' - Log100}{Log40000 - Log100}\right) = 0.0167 \tag{2.12}$$

or

$$\frac{1}{2.602}Log\frac{X}{X'} = 0.0167 \tag{2.13}$$

The final expression is given by:

$$\frac{X}{X'} = 10^{0.04345} = 1.105 \tag{2.14}$$

This increase of GDP per-capita rate compensates a one-year decrease of life expectancy.

For example, if the reference income (X') is 100 \$, then a one-year decrease of life expectancy can be counterbalanced by a higher income, corresponding to $X'' = 110.5$ \$.

In general, a n-years decrease of life expectancy can be compensated by the following GDP per-capita increase:

$$X = X' \cdot 10^{n \cdot 0.04345} \tag{2.15}$$

The term $10^{n \cdot 0.04345}$ is the "substitution rate" between a n-years life expectancy decrease and a corresponding GDP increase.

The increase in GDP depends on the GDP reference value [3] (X'):

$$X - X' = X' \cdot 10^{0.04345} - X' = X' \left(10^{0.04345} - 1\right) = 0.105 \cdot X' \qquad (2.16)$$

To compensate a one-year life expectancy decrease, higher GDP reference values (X'), determine higher GDP increases. For example, Congo GDP per-capita is 970 \$ (UNDP 2003), then:

$$\Delta X = X - X' = 0.105 \cdot X' = 101.85 \text{ \$} \qquad (2.17)$$

Spain GDP per-capita is 20150 \$ (UNDP 2003), then $\Delta X = 2115.75$ \$.

The life expectancy for poorer countries is counterbalanced by lower GDP values than for richer countries. Extending this idea to the limit, the life expectancy for richer countries is increased more than the one for poorer countries!?!

Other substitution rates can derive from Eq. 2.8. For example the relation between LEI and ALI:

$$\Delta(\text{LEI}) = -\frac{2}{3}\Delta(\text{ALI}) \qquad (2.18)$$

In general, a one-year decrease of life expectancy can be compensated by an ALI increase, corresponding to:

$$\left|\Delta(\text{ALI})\right| = \frac{3}{2}\left|\Delta(\text{LEI})\right| = \frac{3}{2}0.0167 = 0.025 \qquad (2.19)$$

Equally, to compensate a n-years life expectancy decrease, the adult literacy index (ALI) has to increase by a $n \cdot 0.025$ factor.

---

[3] Differentiating Eq. 2.11: $d(\text{LEI}) = d\left(\dfrac{LogX - Log100}{Log40000 - Log100}\right)$

from which: $d(\text{LEI}) = \dfrac{1}{2.6} \cdot d\left[Log\left(X/100\right)\right]$ then:

$$d(\text{LEI}) = \frac{1}{2.6} \cdot \frac{1}{X/100} \cdot \frac{1}{100} \cdot Log_{10}e \cdot dX$$

and simplifying: $d(\text{LEI}) = 0.1669\dfrac{dX}{X}$ or $dX = X \cdot \dfrac{d(\text{LEI})}{0.1669}$.

Since $d(LEI) = 0.0167$, then: $dX = 0.105 \cdot X$, which is the same of Eq. 2.16

Indicators independence

Let consider two countries, X and Y, which have the same life expectancy value (which is rather weak), but the adult literacy index of Y is lower than that one of X, and Y exceeds X with regard on GDP (see Table 2.4) (Bouyssou et al. 2000).

Since life expectancy is extremely low (and consequently there are very few adults) ALI can be seen as a factor of little importance, as opposed to the GDP per-capita. In fact, it is reasonable to suppose that a GDP increase, leads to health conditions and life expectancy enhancements. In conclusion, Y can be considered more developed than X. The HDI values (0.345 for X and 0.367 for Y) confirm this assertion.

Table 2.4. Property of independence of the indicators. Values of life expectancy, ALI, ERI and GDP per-capita for two generic countries (X and Y). With regard on HDI, Y exceeds X

Country	LEI	ALI	ERI	GDP per-capita	HDI
X	30	0.70	0.65	500	0.345
Y	30	0.35	0.40	4 000	0.367

Let us consider two other countries – W and Z – whose indicators are similar to X and Y, except for life expectancy which is 70 for both W and Z (see Table 2.5). The Z's ALI value is half of the W's. In this case, while the adult population is quite large (life expectancy is 70), Z has an important problem of illiteracy. Even if the high Z's yearly income (high GDP per-capita) will concur to enhance the diffusion of literacy, this diffusion may require several years. On the other hand, the low GDP per-capita of W is not such a major problem regarding the quality of life, since the level of life expectancy and adult literacy is quite high. It seems reasonable to conclude that W is more developed than Z.

In spite of all, this HDI value is 0.567 for W, and 0.589 for Z! This should not be surprising, considering that there is no difference between the pair X and Y, and the pair W and Z, except for life expectancy.

The difference in life expectancy between X and W, is the same as the one between Y and Z. For that reason, the HDIs of W and Z equally increase, respectively with regard to those of X and Y. In this precise situation, for the same values of ALI, ERI and GDPI, HDI is directly proportional to LEI. Changes in the LEI directly impact on the HDI.

$$\Delta(HDI) = \frac{1}{3}\Delta(LEI) \tag{2.20}$$

Table 2.5. Property of independence of indicators. Values of life expectancy, ALI, ERI and GDP per-capita for two generic countries (W and Z). With regard to HDI, Z exceeds W. The result is consistent with the values in Table 2.4, even if it is contradictory with the hypothesis of development for a country

Country	LEI	ALI	ERI	GDP per-capita	HDI
W	70	0.70	0.65	500	0.567
Z	70	0.35	0.40	4 000	0.589

When different "dimensions" of a model are aggregated by a sum operator, the identical performances of two countries, with regard to one or more dimensions, are not relevant for their comparison. These equal dimensions can be changed in value, with no influence on the comparison result. This characteristics is known as *"dimensional independence"*, and – as the example shows – sometimes it can be undesirable.

Indicators scale construction

The concept of scale construction has already been introduced in Sect. 2.5.5. In this paragraph we focus on other important aspects.

Let us consider, for example, GDPI (Eq. 2.6). The GDPI is adjusted using a logarithmic function. This is not the only possibility. Until recently, the calculation of the income per-capita was performed using the Atkinsons' algorithm as follows (Atkinsons 1970):

$$W(y) = \begin{cases} y & \text{if } 0 < y < y^*, \\ y^* + 2\left[(y - y^*)^{1/2}\right] & \text{if } y^* \le y < 2y^*, \\ y^* + 2(y^*)^{1/2} + 3\left[(y - 2y^*)^{1/3}\right] & \text{if } 2y^* \le y < 3y^*, \\ \qquad\qquad ..... \\ y^* + 2(y^*)^{1/2} + 3(y^*)^{1/3} + ... + n\left[(y - (n-1)y^*)^{1/n}\right] & \text{if } (n-1)y^* \le y < ny^* \end{cases} \qquad (2.21)$$

where:
y income per-capita;
$W(y)$ transformed income;
y^* annual world average income (for example, in 1994 $y^* = 5835$ \$ (UNDP 1997)). y^* depends on the examined year.

The GDPI value was calculated using the following formula:

$$\text{GDPI} = \frac{W(\text{GDP per-capita}) - W(100)}{W(40000) - W(100)} \qquad (2.22)$$

With reference to the year 1994, $W(40000) = 6154$ and $W(100) = 100$.

Comparing the results obtained by each method – Eqs. 2.6 and 2.22 – we can notice significant differences. For example, in 1994 Greece obtained a 11265 \$ income per-capita; using the Atkinsons' algorithm we obtain:

$$\text{GDPI}_{Greece-1994} = \frac{W(\text{GDP per-capita}) - W(100)}{W(40000) - W(100)} = \frac{5982 - 100}{6154 - 100} = 0.972 \quad (2.23)$$

The value of 5982 derives from the expression:

$$5982 = 5835 + 2\left[(11265 - 5835)^{1/2} \right] \quad (2^{nd} \text{ line of Eq. 2.21}). \quad (2.24)$$

On the other hand, when using Eq. 2.6 we obtain:

$$\text{GDPI}_{Greece-1994} = \frac{Log\ 11265 - Log\ 100}{Log\ 40000 - Log\ 100} = 0.789 \quad (2.25)$$

Algorithms like Atkinsons' (Eq. 2.21) depend on parameters which change every year (value of y^*). As a consequence, comparability among different years is quite difficult. The first method proposed (Eq. 2.6) is not affected by this problem.

It is important to remark that the choice of different adjusting algorithms, even if they are equally defensible from an economic point of view, produces different evaluations in terms of human development.

Other adjusting functions, as well as that proposed for GDP, could have been used to adjust the life expectancy value, the adult literacy rate, and so on. For example, a one-year increase of life expectancy is more relevant for a country where LEI is 30 years, than for a country where LEI is 70 years.

Statistical remarks on HDI

The life expectancy indicator is calculated as the mean value of an entire population. In general, the mean value, even if it provides useful information, does not consider the composition of the examined population. In a country where all the people live up to 50 years, the life expectancy is 50 years. In a country where half of the population lives up to 70 years and half up to 30 years, the life expectancy is 50 years. However, the two conditions are very different (Bouyssou et al. 2000).

The mean value of life expectancy is very different from the mean value of the size or the weight of some objects. In fact, weight is an observable manifestation of an object, whereas the mean value of life expectancy does not correspond to anything real. It is just a convention. It is the life length of an ideal (average) human.

With reference to the HDI calculation, it is important to think over the meaning of each of the four sub-indexes. Are they all mean values? What is the correct interpretation of the aggregated (sub)indicators? These questions typically refer to all the indicators synthesising different dimensions of a complex system.

2.3 Air quality indicators

This section analyses and compares three existing Indicators of the air quality: the American AQI, the French ATMO, and the Italian IQA.

As a consequence of the frightening increase in air pollution, especially in large urban areas, many international organizations and governments have adopted a series of regulations to keep a low the level of pollutant concentration. Since people incessantly breathe in the atmosphere, the presence of pollutants is very dangerous for human health.

For several years, a thorough scientific research has been carried out in order to find out how much the pollutants diffused in the air affect people. This activity is carried out monitoring chemically and physically the environment and also using biological indicators to evaluate the impact the pollutants have on the ecosystem and on people (Rapport et al. 2003). For instance, several studies established that nitrogen dioxide (NO_2) and ozone (O_3) increase the risk of death in patients with severe asthma (Sunyer et al. 2002). Ozone polluting the air generates an increase in lung cancer (Yang et al. 2005). Traffic-related air pollution increase mortality (Hoek et al. 2002). Ozone and carbon monoxide (CO) are linked to cardiac birth deficiencies (Ritz et al. 2002) etc.

Public awareness of this problem has been increasing considerably in the past few years due to the media. A high level of air pollution is not only harmful and annoying to the population, but is also a heavy drain on the wealth of a Country. The health damages generate several additional charges relating to health service, mobility and absence from school or work due to sickness, monitoring and protection of the environment.

For this purpose, several countries have introduced evaluation methods that rapidly and efficiently indicate the air quality condition for the population. First, the United States *Environmental Protection Agency* (EPA) developed the Air Quality Index - AQI (United States Environmental Protection Agency, 1999). This indicator provides air quality information using numeric and chromatic indicators, which allow an immediate and extensive evaluation of risk for the human health.

Other European countries have followed the way. The following sections illustrate the ATMO index - formulated and made operational by the French Environment Ministry (Ministère de l'écologie et du développement durable, 2004) - and the IQA index - developed and made operative in some northern regions of Italy: Piedmont, Lombardy, etc.. (Piedmont Regional Law43/2000 2000). These indicators are similar in pattern, but are operated in different ways. Briefly, they take into account the concentrations of the main air pollutants (usually measured by µg/m^3), and set a tolerability scale for each pollutant, trying to gather general information on the overall air condition.

2.3.1 The American Air Quality Index (AQI)

The AQI is used for the Metropolitan Statistical Areas (MSAs) of US, with a population of more than 350,000 – according to the *Clean Air Act* safety limits of five air pollutants: Ozone (O_3), Particulate Matter (PM), Carbon Monoxide (CO), Sulphur Dioxide (SO_2), Nitrogen Dioxide (NO_2) (US E.P.A., 1999).

In each Area, the previous 24 hours concentration of the five mentioned pollutants are measured (or estimated) and reported on six reference categories (Table 2.6).

Table 2.6. The six reference categories of the five pollutants embraced by the AQI (Air Quality Index). The calculation method and breakpoints of the reference categories are specified for every pollutant (US E.P.A. 1999)

O_3 8-hour mean value [µg/m^3]	PM_{10} 24-hour mean value [µg/m^3]	CO 8-hour mean value [µg/m^3]	SO_2 24-hour mean value [µg/m^3]	NO_2 1-hour mean value [µg/m^3]	AQI reference values
0–137	0–54	0–5.5	0–97	(*)	0–50
138–180	55–154	5.6–11.76	98–412	(*)	51–100
181–223	155–254	11.77–15.5	413–640	(*)	101–150
224–266	255–354	15.6–19.25	641–869	(*)	151–200
267–800	355–424	19.26–38.0	870–1727	1330–2542	201–300
> 800	425–604	38.1–50.5	1728–2300	2543–4182	301–500

(*) CE regulations do not set a Nitrogen Dioxide (NO_2) short-term limit. They only define a yearly mean value of 100 µg/m^3. In the AQI calculation the Nitrogen Dioxide is considered uniquely if the hourly mean concentration is higher than 1330 µg/m^3 (so according to AQI reference values upper than 200).

According to the *Clean Air Act*, an AQI value of 100 or less represents an acceptable concentration for each pollutant. Therefore, AQI values lower than 100 are judged admissible. A higher AQI value means that the air is unhealthy for the more sensitive subjects, and as the air pollution increases it also becomes unhealthy for the general population.

It's interesting to notice that the above-mentioned thresholds are considerably higher than the EU (European Union) regulations. The EU limits are respectively 120 $\mu g/m^3$ for O_3, 150 $\mu g/m^3$ for PM_{10} (24-hour mean value), and 100 $\mu g/m^3$ for NO_2 (yearly mean value) (European Community, Dir. 96/62/EC, 2002/3/EC).

For each area, the daily AQI value is given by the worst registered condition among the five pollutants:

$$AQI = \max\left\{ I_{O_3}, I_{PM_{10}}, I_{CO}, I_{SO_2}, I_{NO_2} \right\} \tag{2.26}$$

The more the AQI value grows, the higher the health risk is.

Given the pollutants' concentration data and the breakpoints in Table 2.6, every AQI sub-index is calculated using Eq. 2.27 (linear interpolation):

$$I_p = I_{L,p} + \frac{I_{H,p} - I_{L,p}}{BP_{H,p} - BP_{L,p}} \left(C_p - BP_{L,p} \right) \tag{2.27}$$

where:

I_p the sub-index for the pth pollutant;
C_p the concentration of the pth pollutant;
$BP_{H,p}$ the breakpoint that is greater than C_p;
$BP_{L,p}$ the breakpoint that is less than or equal to C_p;
$I_{H,p}$ the AQI value corresponding to $BP_{H,p}$;
$I_{L,p}$ the AQI value corresponding to $BP_{L,p}$.

For instance suppose you have an 8-hour ozone (O_3) concentration of 187 $\mu g/m^3$. Then you look in Table 2.6 for the range that contains this concentration (first column, third row), which is placed between the breakdowns $BP_{L,O_3} = 181 \mu g/m^3$ and $BP_{H,O_3} = 223 \mu g/m^3$, corresponds to the sub-index values of $I_{L,O_3} = 101$ to $I_{H,O_3} = 150$. So an ozone concentration of 187 $\mu g/m^3$ corresponds to a sub-index given by Eq. 2.28:

$$I_{O_3} = I_{L,O_3} + \frac{I_{H,O_3} - I_{L,O_3}}{BP_{H,O_3} - BP_{L,O_3}}\left(C_{O_3} - BP_{L,O_3}\right)$$

$$= 101 + \frac{150 - 101}{223 - 181}(187 - 181) = 108 \tag{2.28}$$

Now consider the air condition in Table 2.7.

Table 2.7. Air pollutant values registered in a particular Metropolitan Area

Registered Values			
PM_{10} $[\mu g/m^3]$	O_3 $[\mu g/m^3]$	CO $[\mu g/m^3]$	SO_2 $[\mu g/m^3]$
158	165	10.5	66

$$I_{PM_{10}} = I_{L,PM_{10}} + \frac{I_{H,PM_{10}} - I_{L,PM_{10}}}{BP_{H,PM_{10}} - BP_{L,PM_{10}}}\left(C_{PM_{10}} - BP_{L,PM_{10}}\right)$$

$$= 101 + \frac{150 - 101}{254 - 155}(158 - 155) = 102 \tag{2.29}$$

$$I_{O_3} = I_{L,O_3} + \frac{I_{H,O_3} - I_{L,O_3}}{BP_{H,O_3} - BP_{L,O_3}}\left(C_{O_3} - BP_{L,O_3}\right)$$

$$= 51 + \frac{100 - 51}{180 - 138}(165 - 138) = 82 \tag{2.30}$$

$$I_{CO} = I_{L,CO} + \frac{I_{H,CO} - I_{L,CO}}{BP_{H,CO} - BP_{L,CO}}\left(C_{CO} - BP_{L,CO}\right)$$

$$= 51 + \frac{100 - 51}{11.76 - 5.6}(10.5 - 5.6) = 90 \tag{2.31}$$

$$I_{SO_2} = I_{L,SO_2} + \frac{I_{H,SO_2} - I_{L,SO_2}}{BP_{H,SO_2} - BP_{L,SO_2}}\left(C_{SO_2} - BP_{L,SO_2}\right)$$

$$= 0 + \frac{50 - 0}{97 - 0}(66 - 0) = 34 \tag{2.32}$$

The AQI is 102 with PM_{10} as the responsible pollutant.

$$AQI = \max\left\{I_{PM_{10}}, I_{O_3}, I_{CO}, I_{SO_2}\right\} = \max\left\{102, 82, 90, 34\right\} = 102 \tag{2.33}$$

Each AQI value is linked with a colour and with an air quality descriptor. The AQI scale is split into six reference categories by E.P.A. - the same reported in Table 2.6. The more the AQI value increases, the more the population health risk increases (see Table 2.8)

For example, when the AQI is 50, the air quality is good with a low risk level, and the associated colour is green. Vice-versa for an AQI higher than 300, the air quality is bad with a high risk level, and the associated colour is maroon.

Table 2.8. American AQI Categories, Descriptors, and Colours (US E.P.A. 1999)

AQI Reference Values	Descriptor	Colour
0–50	Good	Green
51–100	Moderate	Yellow
101–150	Unhealthy for sensitive groups	Orange
151–200	Unhealthy	Red
201–300	Very unhealthy	Purple
> 301	Hazardous	Maroon

The E.P.A. qualitative description related to the AQI categories are as follows:

- *"Good"*: the AQI value is within the 0-50 range. The air quality is satisfactory, with very little risk to the population.
- *"Moderate"*: AQI included between 51 and 100. The air quality is admissible, however a few people could be healthy damaged because of the presence of pollutant. For instance, ozone sensitive people may experience respiratory symptoms.
- *"Unhealthy for sensitive groups"*: children and adults with respiratory disease are at risk when doing outdoor activities, due to the ozone exposure, whereas people with cardiovascular disease are most at risk due to the carbon monoxide exposure. When the AQI value is included between 101 and 150, these sensitive individuals could increase their symptoms of disease to the point of health compromising. However, much of the population is not at risk.
- Within the 151-200 range the AQI is considered to be *"unhealthy"*. This situation causes possible disease for the general population. Sensitive individuals could seriously suffer.
- *"Very unhealthy"* AQI values – between 201 and 300 – represent an alarm. The whole population could be health damaged seriously.
- *"Hazardous"* – over 300 – AQI values trigger an immediate alarm. The whole population is at serious risk of suffering from diseases.

This index puts the emphasis of the risk on the groups most sensitive to air pollutant like children, elderly people, and people with respiratory or cardiovascular disease. Adequate communications support the AQI utilization and have made it available to population. The colour format – which represents the listed categories, the pollution level and the linked health risk – allows people to react to their existing circumstances.

The AQI property of non-monotony

In reference to the example data in Table 2.7, let us suppose the Carbon Monoxide concentration increases, driving the corresponding index I_{CO} to 102. This new condition is surely worse than the previous one (two of the four sub-indexes have a value of 102). Nevertheless, the AQI indicator remains unchanged and it does not properly represent significant air pollution changes. Therefore, the AQI does not fulfil the property of monotony.

The AQI property of non-compensation

Suppose we were to calculate the AQI taking into account the two different air quality conditions (W and Z) in Table 2.9.

The first set of data (W) is almost perfect except for the PM_{10} concentration, where as the second one (Z) is not particularly good in all the sub-indexes. Nevertheless, the AQI is 155 in the situation (W) and it is 130 in the situation (Z). Unlike other indicators, the AQI does not fulfil any sub-index compensation.

Table 2.9. AQI sub-indexes values in two air quality conditions. Although some pollutant concentrations (O_3, CO and SO_2) are worse in condition Z, the AQI is higher in condition W

Condition	PM_{10}	O_3	CO	SO_2
(W)	155	30	25	30
(Z)	130	104	100	121

The AQI sub-indexes scale construction

In regard to Table 2.6 it clearly appears that a "homogeneous mapping" - between each pollutant concentration bandwidth and the corresponding sub-index variation size - is lacking.

Considering, for example, the SO_2 pollutant, you notice the first AQI level relates the [0–97 $\mu g/m^3$] range of concentration and the [0–50] sub-index range; otherwise the second AQI level link is between the [98–412 $\mu g/m^3$] range of concentration and the [51–100] sub-index range. The AQI level range width is constant, but the width of the range of concentration differs.

Each pollutant range size is fixed on the basis of the pollutant concentration effects on human health. This assumption partly contrasts with the direct proportionality assumption between the air pollutant concentration and the AQI within each specific range (see Eq. 2.27).

2.3.2 The ATMO index

The ATMO index was developed by the French Environment Ministry. It is based on the concentration of four air pollutants: Ozone (O_3), Particulate Matter (PM), Sulphur Dioxide (SO_2), and Nitrogen Dioxide (NO_2) (Ministère de l'écologie et du développement durable 2004; Bouyssou et al. 2000). Each of the pollutants is related to a sub-index. Each pollutant concentration is measured and reported on a ten level scale. The first level corresponds to an excellent air quality, the 5th and 6th level are just around the European long-term norms, the 8th level corresponds to the EU short-term norms, and finally the 10th level corresponds to a health hazard condition. The sub-indexes' ten level scales are shown in Table 2.10.

Table 2.10. Ten reference categories of the four sub-indexes which make up the ATMO indicator. Each level is set out between a minimum and a maximum breakpoint value (Ministère de l'écologie et du développement durable 2004)

Lev.	PM_{10} [$\mu g/m^3$]		O_2 [$\mu g/m^3$]		O_3 [$\mu g/m^3$]		SO_2 [$\mu g/m^3$]		Descriptor	Colour
	Min	Max	Min	Max	Min	Max	Min	Max		
1	0	9	0	29	0	29	0	39	Very Good	Green
2	10	19	30	54	30	54	40	79	Very Good	Green
3	20	29	55	84	55	79	80	119	Good	Green
4	30	39	85	109	80	104	120	159	Good	Green
5	40	49	110	134	105	129	160	199	Medium	Orange
6	50	64	135	164	130	149	200	249	Poor	Orange
7	65	79	165	199	150	179	250	299	Poor	Orange
8	80	99	200	274	180	209	300	399	Bad	Red
9	100	124	275	399	210	239	400	499	Bad	Red
10	≥ 125		≥ 400		≥ 240		≥ 500		Very Bad	Red

The ATMO value is the maximum of the four sub-indexes:

$$\text{ATMO} = \max\left\{ I_{NO_2}, I_{SO_2}, I_{O_3}, I_{PM_{10}}, \right\} \qquad (2.34)$$

The PM_{10} value is the (great) mean of the daily mean values, registered from 1:00 to 24:00 in the different operative stations of the monitored area. The value of the other pollutants is the mean of the maximum hourly values, registered from 1:00 to 24:00 in the different operative stations of the monitored area.

To better describe the ATMO directions for use, consider the air condition in Table 2.11 Coherently with the Eq. 2.34, the ATMO index value is 8. In this specific condition the air quality is very unhealthy.

Table 2.11. ATMO sub-indexes encoding. The ATMO index value is 8

Pollutant	NO_2	SO_2	O_3	PM_{10}
sub-index	2	2	3	8

A brief description about the ATMO main properties is reported in the next sections.

The ATMO property of non-monotony

Let imagine a sunny day with heavy traffic and no wind. In reference to the example data in Table 2.11, let us suppose the Ozone concentration increases driving the corresponding index from 3 to 8. The new condition is surely worse compared to the previous one (two of the four sub-indexes have a value of 8). The ATMO indicator value would be expected to be higher than the previous one. Nevertheless, the ATMO indicator remains unchanged at the value of 8 and it does not properly represent significant air pollution change. This proves that the ATMO does not fulfil the property of monotony.

The ATMO property of non-compensation

Consider the calculation of the ATMO with two different air quality conditions (U and V) in Table 2.12.

Table 2.12. ATMO sub-indexes values in two air quality conditions. Although some Y condition pollutant concentrations (NO_2, SO_2 and PM_{10}) are worse, the ATMO index is higher for the X condition

Condition	NO_2	SO_2	O_3	PM_{10}
(U)	2	1	7	1
(V)	6	5	6	6

The first set of data (U) is almost perfect except for the O_3 concentration, where as the second one (V) is not particularly good for all the sub-indexes. Nevertheless, the ATMO is 7 in the situation (U) and it is 6 in the situation (V).

In this case, the ATMO sub-indexes do not compensate each other. In the state U, the high O_3 value is not counterbalanced by the three remaining sub-indexes' low values.

The ATMO sub-indexes scale construction

The ATMO indicator, as well as the AQI, lacks a "homogeneous mapping" between each pollutant concentration bandwidth and the corresponding sub-index range size (see Table 2.10). For example by considering the SO_2 pollutant, the first ATMO level is related to the [0–39 µg/m^3] range of concentration (bandwidth of 39 µg/m^3), and the sixth ATMO level is related to the [200–249 µg/m^3] range of concentration (bandwidth of 49 µg/m^3). As stated above for the AQI, each ATMO range size is fixed on the basis of the pollutant effects on human health. Every value of the pollutant in the same ATMO reference level is supposed to be equivalent. An SO_2 concentration of 201 µg/m^3 has to be considered equivalent to a 249 µg/m^3 concentration, when looking at the effects on human health.

Incidentally, it's interesting to see that the concentration values, which bound the ATMO reference levels, are considerably smaller than the AQI ones.

2.3.3 The IQA index

In some northern Italian regions (Piedmont, Lombardy, etc.) different systems are currently being tested to monitor the air quality and provide information to the public. The analysis will focus on the IQA index (Indice di Qualità dell'Aria), used in Piedmont (Piedmont Regional Law 43/2000 2000). The index is inspired by the United States *Environmental Protection Agency*, AQI (Sect. 2.3.1), but with some clear differences.

According to the safety regulation limit, the IQA aggregates the most critical air pollutants at each time of the year on the basis of their effects on human health: Ozone (O_3) and Particulate Matter (PM_{10}) in summertime, PM_{10} and Nitrogen Dioxide (NO_2) in wintertime.

The IQA index constitutes a popular communication system that extensively monitors the condition of air quality. The IQA is expressed by a numerical index between 1 and 7; the higher the air pollution is, the more serious the health hazard is and the higher the index value is. Consequently, IQA represents also the human health risk.

The IQA index comprises several sub-indexes, each of which are related to a monitored pollutant: Ozone (O_3), Particulate Matter (PM_{10}) and Nitrogen Dioxide (NO_2). The IQA value is the arithmetic mean of the two maximum sub-indexes' values Eq. 2.35.

$$IQA = \frac{I_1 + I_2}{2} \tag{2.35}$$

where I_1 and I_2 are the two sub-indexes with the higher value (selected among the three critical pollutants PM_{10}, O_3, NO_2).

The IQA is calculated on a daily basis using the values of the concentration of pollutants from the previous 24 hours. The IQA numerical value is converted into a 7 level reference scale, featured in Table 2.13. This final class rank is presented to the population. This information is enriched by an indication of the air pollution evolution, derived from the weather forecast.

Table 2.13. IQA Categories, Descriptors and Colours (Piedmont Regional law 43/2000 2000)

IQA reference values	IQA final rank	Descriptor	Colour
0-50	1	Excellent	Blue
51-75	2	Good	Light blue
76-100	3	Fair	Green
101-125	4	Mediocre	Yellow
126-150	5	Not very healthy	Orange
151-175	6	Unhealthy	Red
>175	7	Very unhealthy	Purple

Briefly, IQA is a conventional indicator used to:

- report the quality of the air on a daily basis;
- identify the worst environmental parameters;
- calculate the amount of risk that the population is subjected to.

A numeric value of IQA equal to 100 essentially corresponds to the air quality safety level for polluting substances. IQA values less than 100 are generally satisfactory with no potential hazard for public health. The more the IQA value is over 100, the more the air quality is considered unhealthy - initially only for the most sensitive groups of people and then for all the population.

Different descriptions of the air quality, different colours, and some useful advice for the population are associated with each of the seven IQA levels:

- "*Excellent*" – blue, with a numeric IQA value between 0 and 50. The quality of the air is considered excellent.
- "*Good*" – light blue, with a numeric IQA value between 51 and 75. The air quality is considered very satisfactory with no risk for the population.
- "*Fair*" – green, with a numeric IQA value between 76 and 100. The air quality is satisfactory and there is no risk for the population.
- "*Mediocre*" – yellow, with a numeric IQA value between 101 and 125. The population is not at risk. People with asthma, chronic bronchitis or heart problems might show symptoms of slight breathing problems, but only during intense physical activity; it is advised that people with these ailments category of people limit their physical exercise outdoors, especially during the summertime.
- "*Not very healthy*" – orange, with a numeric IQA value between 126 and 150. People with heart problems, the elderly and children may be at risk. It is advised that these categories of people limit their physical activity and prolonged periods of time outdoors, especially during the peak daytime hours in summertime.
- "*Unhealthy*" – red, with a numeric IQA value between 151 and 175. Many people could have slightly negative health problems, albeit reversible; it is advised to limit extended periods of time outdoors, especially in the peak daytime hours during the summertime. People in the sensitive groups could, however, may have more serious symptoms; in these cases it is highly recommended expose oneself as little as possible to the open air.
- "*Very unhealthy*" – purple, with a numeric IQA value above 175. There may be slightly negative effects on the health of all people in the area. The elderly and people with respiratory problems (breathing difficulties) should avoid going outside. Other people (especially children) should avoid doing physical activity and limit their time outdoors, especially during the peak daytime hours in summertime.

The IQA sub-indexes are:

- *Nitrogen dioxide* (NO_2) Sub-index: I_{NO_2}

$$I_{NO_2} = \frac{\bar{V}_{\max h_{NO_2}}}{V_{rif\, h_{NO_2}}} \cdot 100 \qquad (2.36)$$

$\overline{V}_{\max h_{NO_2}}$ is the mean of the maximum hourly NO_2 concentration values, registered from 1:00 to 24:00 in the different operative stations of the monitored area;

$V_{rif h_{NO_2}}$ is a NO_2 concentration reference value (200 µg/m³), which represents a hourly safety limit (Ministero dell'ambiente e della tutela del territorio 2002).

- *Particulate matter* (PM$_{10}$) Sub-index: $I_{PM_{10}}$

$$I_{PM_{10}} = \frac{\overline{V}_{med\ 24h_{PM_{10}}}}{V_{rif\ PM_{10}}} \cdot 100 \qquad (2.37)$$

$\overline{V}_{med\ 24h_{PM_{10}}}$ is the arithmetic mean of the average hourly PM$_{10}$ concentration values, registered from 1:00 to 24:00 in the different operative stations of the monitored area;

$V_{rif\ PM_{10}}$ is a PM$_{10}$ concentration reference value (50 µg/m³), which represents a daily safety limit (Ministero dell'ambiente e della tutela del territorio 2002).

- *Ozone* (O$_3$) Sub-index: I_{8hO_3}

$$I_{8hO_3} = \frac{\overline{V}_{\max\ 8h_{O_3}}}{V_{rif\ 8h_{O_3}}} \cdot 100 \qquad (2.38)$$

$\overline{V}_{\max\ 8h_{O_3}}$ is the mean of the maximum O$_3$ concentration values, registered every 8 hours, and calculated every hour on the basis of the previous 8 hours, from 1:00 to 24:00 in the different operative stations of the monitored area;

$V_{rif\ 8h_{O_3}}$ is a O$_3$ concentration reference value (120 µg/m³), which represents a 8-hours safety limit (Dir. 2002/3/EC).

To allow an evolution's assessment of the atmospheric pollution, the index value of the previous day is reported together with the index values from the six previous days.

As an example, let show the calculation of the IQA index for Turin's metropolitan area for 27 January 2005. The data for each pollutant can be seen below in Table 2.14 (Province of Turin's Regional Agency for the Environment - ARPA 2005).

Table 2.14. Values of pollutants found in Turin's metropolitan area on 27 January 2005 (ARPA, Province of Turin 2005)

Registered Values					
PM$_{10}$	O$_3$	NO$_2$	C$_6$O$_6$	CO	SO$_2$
[μg/m^3]	[μg/m^3]	[μg/m^3]	[μg/m^3]	[μg/m^3]	[μg/m^3]
84	33	125	5.6	2.2	15

The first three pollutants are used to calculate the IQA value. The associated sub-indexes are respectively:

$$I_{NO_2} = \frac{125}{200} \cdot 100 = 62.5; \quad I_{PM_{10}} = \frac{84}{50} \cdot 100 = 168;$$

$$I_{8hO_3} = \frac{33}{120} \cdot 100 = 27.5$$

(2.39)

The two maximum values refer to PM$_{10}$ and to NO$_2$. The IQA index value is:

$$\text{IQA} = \frac{I_{NO_2} + I_{PM_{10}}}{2} = \frac{62.5 + 168}{2} = 115.3$$

(2.40)

This value corresponds to level 4 (mediocre) according to the IQA scale indicated in Table 2.13.

The IQA property of non-monotony

The IQA indicator is not monotonous. Taking into consideration the data in Table 2.14, one supposes that the sub-index "ozone" goes from 27.5 to 61. This latter condition (obviously worse than the previous one) is still described by the same IQA indicator value (115.3). Also in this case the IQA indicator does not adequately evaluate significant changes in the concentration of pollutants.

The IQA property of compensation

Let us calculate the IQA considering the two different air quality conditions (X and Y) in Table 2.15.

The IQA associated sub-indexes' values are respectively:

$$I_{NO_2} = \frac{125}{200} \cdot 100 = 62.5; \quad I_{PM_{10}} = \frac{84}{50} \cdot 100 = 168;$$

$$I_{8hO_3} = \frac{33}{120} \cdot 100 = 27.5 \tag{2.41}$$

$$I_{NO_2} = \frac{336}{200} \cdot 100 = 168; \quad I_{PM_{10}} = \frac{31.25}{50} \cdot 100 = 62.5;$$

$$I_{8hO_3} = \frac{33}{120} \cdot 100 = 27.5 \tag{2.42}$$

Table 2.15. Concentrations of pollutants in two air quality conditions. Although the concentrations are different, the IQA index value is the same

	Registered Values		
Condition	PM_{10} [$\mu g/m^3$]	O_3 [$\mu g/m^3$]	NO_2 [$\mu g/m^3$]
(X)	84	33	125
(Y)	31.25	33	336

In both conditions, the two maximum values refer to the PM_{10} and NO_2 sub-indexes, and drive the IQA indicator to the same value:

$$\text{IQA} = \frac{I_{NO_2} + I_{PM_{10}}}{2} = \frac{62.5 + 168}{2} = 115.3 \tag{2.43}$$

Based on the health risk estimated by the IQA index, the two previous conditions are considered equivalent. We can define a sort of "substitution rate" (Eq. 2.35):

$$\Delta\left(I_{NO_2}\right) = -\Delta\left(I_{PM_{10}}\right) \tag{2.44}$$

in terms of pollutants' physical concentration (apply Eq. 2.36 and Eq. 2.37):

$$\frac{\Delta C_{NO_2} \cdot 100}{200} = -\frac{\Delta C_{PM_{10}} \cdot 100}{50} \tag{2.45}$$

where:

ΔC_{NO_2} variation in the NO_2 concentration

$\Delta C_{PM_{10}}$ variation in the PM_{10} concentration

from which it follows that:

$$\Delta C_{NO_2} = -4 \cdot \Delta C_{PM_{10}} \qquad (2.46)$$

A 1 $\mu g/m^3$ variation in concentration of PM_{10} is balanced out by a 4 $\mu g/m^3$ variation in concentration of NO_2. Considering the damages to human health, is this "substitution rate" reasonable?

Construction of IQA sub-index scale

According to the linear proportionality formulas given in the Eqs. 2.36, 2.37 and 2.38, there is a uniform "mapping" between the concentrations of the pollutant and the values of the IQA indicator.

The calculation of each sub-index is influenced by the values established in reference to the regulations for the protection of human health (D.M 2.04.2002 n. 60, Dir 2000/3/EC and following daughter directives). Possible changes of reference values may have direct consequences on single part of the sub-indexes and on the possibility to compare them over time. On the other hand, the normative reference values are not clearly mentioned in the definition of the AQI and ATMO indicators.

The Eq. 2.35, which identifies the IQA indicator with the average of the two most critical sub-indexes, is worthy of a special remark. What are the sub-indexes and the IQA's scaling properties? Regarding the possibility of calculating the average, the conventional IQA scale automatically acquires the interval property (Franceschini, 2001). Is that right? The average of the two most critical sub-indexes is really the best model of the problems for the human health?

Finally, is significant to notice that the IQA reference values of PM_{10}, O_3 and NO_2 are considerably lower than the AQI ones.

2.3.4 Comments on indicators meaning

Let us consider the statement: "Today's AQI value is twice (150) as much as yesterday's (75)". Does it make sense? What are the AQI index scaling properties? We will demonstrate the meaninglessness of this statement.

Let us go back to the AQI sub-indexes definition. The concentration of each pollutant is given in $\mu g/m^3$. The conversion into the AQI scale (Table 2.6) is totally arbitrary. For instance, – instead of attributing the values [0–50] to the O_3 pollutant concentrations [0–137 $\mu g/m^3$], and the values [51–100] to the concentrations [138–180 $\mu g/m^3$]–the values [0–40], [41–80] could have been assigned to them. In both the encoding the AQI indicator preserves the same properties.

The most significant information the indicators provide is not the encoded value itself, but the capacity to clarify the tolerability conditions of the air pollution, according to the health risk official standards. The six level encoding is only a way to make the information practical for the user. The AQI sub-indexes scales have only the ordering property. The values of the pollutants concentrations are tied to a level digit, without any rule of linear proportionality. Therefore, the statement "Today the ozone sub-index is higher than yesterday" is reasonable (Franceschini et al. 2005).

On the other hand, the statement: "Today the AQI index is higher than yesterday's" has to be considered with attention. The AQI overall indicator does not fulfil the property of monotony (see Sect. 3.2). Some questions arise when different sub-indexes produce the same AQI index value. An AQI value [51–100] due to the SO_2 sub-index can be compared with the same value due to the PM_{10} sub-index to what extent? What are the risks of the pollutant interactions on health?

Of course competent authorities have carefully built the sub-indexes' scales, considering the single pollutant effects on health. For instance, each sub-index interval [51–100] is linked to the specific pollutant concentration that is tolerated in the American standards. A question on the concept of equivalence is still open. Some pollutants could damage the human health in the short term, others requires a longer time. The effect can be different on the different parts of the body, and so on.

In conclusion, how can probable diseases be estimated? …through physical damages; … health care cost; …the mortality rate…?

2.3.5 Air quality indicators comparison

In the previous sections the paper has indicated some similarity but also substantial differences, among the three air quality indicators: AQI, ATMO, IQA. Although the considered pollutants are similar (PM_{10}, O_3, NO_2, SO_2…), generally the indicators are differentiated by:

- the calculation of the sub-indexes;
- the number of classes of risk;
- the reference categories of the concentrations;
- the criteria of synthesis of sub-indexes;
- the sub-indexes' compensation property.

These distinctions emphasize how the same physical phenomenon can be represented in different ways. Nevertheless, with reference to the concentration of the same pollutants, should the health risk be the same? The following chapters will provide an organic discussion on the problem of the *uniqueness of representation*.

A common aspect of the three discussed indicators is that, for each definite value, the main responsible pollutant is not made explicit. For example, if the ATMO index is 6, we cannot know which is the most critical sub-index responsible for this result. The same goes for AQI and IQA.

It is significant to confirm that indicators are not measurements, although they are defined on the basis of physical values (see Chaps. 3 and 6). Given a particular combination of air pollutants, each of the three indicators leads to a different model. Each indicator maps the pollutants' real concentrations into a different scale, whose level number and range values are absolutely conventional (see Chap. 3).

Another significant aspect of the three indicators is that the environmental conditions, which they are connected to, can be much different. For instance, if the ATMO is 8, the most critical sub-index may possibly be either SO_2 or PM_{10}. Several different concentrations of the air pollutants are supposed to be equally dangerous for the human health (see Table 2.10). In other terms, overall comparable indexes don't entail comparable physical concentrations of the air pollutants.

2.4   The Decathlon competition

Indicators play an important role in sport competitions. In many sports, specific indicators determine the final ranking of single competition or entire championship. Let us consider, for example, formula one races, tennis competitions, artistic gymnastics, synchronized swimming, etc...(Lins et al. 2003; Bouyssou et al. 2000). In particular, it is interesting to analyse the scoring method related to Decathlon competition.

Decathlon is an athletic competition containing 10 different tracks and field (athletics) contests and the winner is the participant (men only) which amasses the highest overall score. Decathlon is a two-day miniature track meet, designed to ascertain the best all-around athlete. Within its competitive rules, each athlete must sprint for 100 meters, long jump, heave a 16-pound shotput, high jump and run 400 meters − all in that very order − on the first day. On the second day the athlete runs a 110 meter hurdle race over 42 inch barriers, hurls the discus, pole vaults, tosses a javelin and, at the end of the contest, races over 1500 meters, virtually a mile.

Decathlon was first introduced at the 1912 Olympic Games of Stockholm, as a three-day multi-event contest. At the begin, the score of the single event was given by the order of arrival. For example, if an athlete finished third in a particular event, he gained 3 points. The winner of the competition was the athlete with the lowest overall score (Zarnowsky 1989). This approach is close to the scores aggregation method presented by Borda (see Chap. 3) (Vansnick 1986; Roy 1996). The main drawback of this method is that it overlooks the athletes' performance level. An athlete who arrives 0.1, 1, or 10 seconds before the next one, finishing a contest in i-th position, always gains i points, and $i+1$ the athlete after him, independently of the performance level.

The problem has been solved introducing some merit scoring tables, connecting the absolute result of each event with a specific score (for instance the time of 10.00 seconds in the 100 metres is actually worth 1096 points). Unlike the first scoring method (based on the order of arrival), in this case the higher the score the better the performance. So, the winner of the competition is the athlete who has scored the highest number of points.

The scoring tables have been published in different versions over the years. At first, the tables' construction was based on the idea that when the performance is close to the world record, it is more difficult to improve it. Therefore, as the performance increases, the score increases more than proportionally.

Fig. 2.2 shows a graphic representation ("convex" curve) of the scoring tables initially suggested by IAAF (International Association of Athletics Federations). The example refers to the distance based events. For the time based events, the graph shape is different: as time spent decreases, the score increases proportionally.

But the "convex" scoring tables generates problems. If one athlete specializes in a subset of events (for example 4 events), neglecting the others, he may have an advantage over the other athletes. By obtaining results close to the world records in his favourite events, he will score many points – of course – at the cost of penalizing the other events. Poor scores in unfavourable contests are largely compensated by high scores in the favourable ones (due to the higher curve slope; see Fig. 2.2). Consequently, it is more convenient to specialize in few contents, rather than preparing for them all with the same commitment. But, this is not in the spirit of decathlon, which consists in recompensing athletes eclecticism, or the ability of performing well is the major part of the events.

To prevent this problem, in 1962 the International Olympic Committee suggested new scoring tables, which can be graphically represented as curves with concave profile (Fig. 2.3). The purpose is to discourage poor performances, stimulating athletes' commitment in all the ten events.

Since 1962, scoring curves have been periodically refined. By way of example, Fig. 2.4 (a) and (b), shows the scoring curves for the high jump and the 100 metres, updated on 2005 (IAAF 2005).

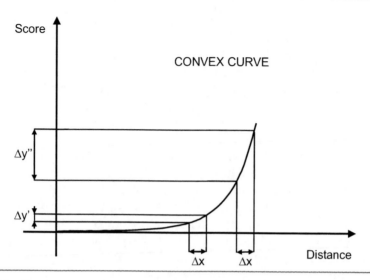

Fig. 2.2. Graphic representation of the decathlon scoring tables (referred to the distance based events) used over the period 1934-1962. When the performance is close to the world record, the score increases more than directly proportionally (convex curve)

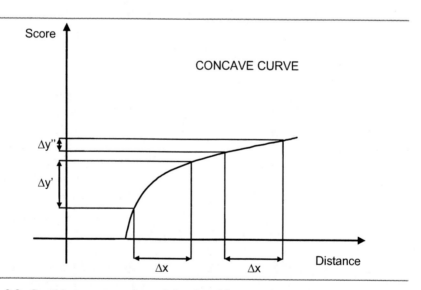

Fig. 2.3. Graphic representation of the decathlon scoring tables (referred to the distance based events) used in the period after 1962. The purpose of the curve is to discourage poor performances (concave curve)

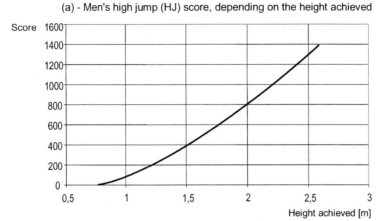

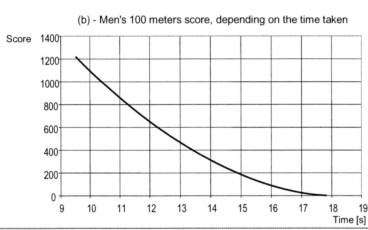

Fig. 2.4. Scoring curves for the high jump (a) and the 100 meters (b) (IAAF 2005)

The analysis of the decathlon scoring tables stimulates the following considerations (Bouyssou et al. 2000):

- How to choose performance limit values (minimum and maximum) of the scoring tables? As already discussed for HDI, the choice of these reference values has direct consequences on the total score calculation (see Sect. 2.2.5). Furthermore, since the maximum value depends on the event's world record, the scoring tables are inevitably subject to change over time.

- Why use an additive rule to aggregate scores in different events? For example, why cannot we use a multiplicative aggregation rule? A multiplicative relation could be in favour of those athletes which get a good score in all the contests. To better explain the concept, let consider a triathlon (3 contents) competition, where each content is associated to a $0 \div 10$ score. One athlete (X) gets 8 points in all the three events, while another one (Y) gets the scores 9, 8 and 7. If we use an additive rule, both athletes totalize 24 points. If we use a multiplicative rule, X gets 512 points, and Y 504 points only. In this case, the athlete with high performances in all the contents is recompensed – as in the spirit of decathlon.

2.4.1 The effects of scoring indicators

Although the main function of decathlon scoring rules is to determine the winner of a competition, they have direct effects on the race strategies. We have analyzed the most significant ones.

First of all, many members of the staff use the scoring indicators to evaluate the athletes' performances. What matters is not the performance level, but the related score. When one athlete gets a score close to the winner's, he is considered a good athlete. This statement is basically true, but it should be analysed in a more detail. The overall score "hides" the separate contributions of the single events. Low scores can be due to athletes who are outstanding in their stronger events, but inadequate in other ones.

A second aspect concerns athletes' preparation. Scores can influence preparation strategies and the definition of the medium-long term objectives. At the beginning, when the final victory was achieved on the basis of the single event placing, the athletes' main goal consisted in overtaking the others. Each race was a separate matter. It was possible to win a single event against mediocre athletes, and losing another event against better ones. A method to estimate the overall performance level was lacking. The scoring tables, introduced later, have changed the importance and the meaning of the single race results.

Athletes not only compete to win races, but also to improve themselves. Apart from the designate winner, the scoring tables are useful tools to compare the results and plan the athletic preparation. In other words, the scoring indicators have become an instrument for monitoring the preparation of the athletes.

These considerations show how indicators, even if defined with the aim of representing the result of a process (decathlon competition in this case), may have unintended and unanticipated consequences to it. The initial representational position of indicators may change, becoming prescriptive.

2.4.2 Representation and decision

In all complex decision making problems, there are several possible criteria for judging the different potential alternatives. The main concern of a Decision Maker is to fulfil his conflicting goals while satisfying the constraints of the system (Vincke 1992; Roy 1993). The literature offers many different approaches for analysing and solving decision making problems.

In the same time, indicators are used to represent the different condition of a process. This representation is fundamental for the process analysis and control (management). The success of the representation depends on the choice of indicators, or rather on how properly they represent the process.

The analogy between decision making models and representational models is evident. Their purpose is to synthesise multi-dimensional information, with the aim of classifying a set of "elements" (see Table 2.16). However, there is a big difference. The construction of indicators – as opposed to the choice of decisional criteria – do not require defining a decisional problem, a decision maker, and a precise set of evaluations.

Table 2.16. Decision and representation: analogies and elements of distinction

Decision	Criteria	Alternatives
Representation	Indicators	Process Conditions

Indicators' field of action is larger than the decision making model. Indicators are not typically used for decision making, but sometimes they can help. The next chapter will discuss the theme of the "non-uniqueness" of the process representation using indicators.

3. The condition of uniqueness in process representation

3.1 Introduction

One of the most critical aspects in operations management is "translating" a firm's goals into performance indicators. The representation is a critical aspect for the process description. "To represent" means transferring the properties of an examined process into a set of indicators, trying to describe the processes' most important aspects.

In the current scientific literature there are many examples of process modeling by means of performance indicators. When dealing, for example, with a given manufacturing plant, indicators such as throughput, defectiveness, output variability, efficiency, etc... are commonly employed (see Chap. 1) (Brown 1996; Kaydos 1999; Galbraith et al. 1991; Maskell 1991; Smith 2000). Special emphasis is also given to the so called "design metrics", that is to say those factors that are inherent in product design, affecting one or more product lifecycle stages (Galbraith and Greene 1995).

Historically, logistics and manufacturing functions are two of the first factory functions to be concerned with the use of performance indicators (Neely et al. 1995; New and Szwejczewski 1995). An interesting survey regarding "logistic metrics" is presented by Caplice and Sheffi (1994). Basing their idea on the conviction that a strategically well designed performance measurement system can be defective at the individual metric level, they state that there is no need for the development of new performance metrics (in logistics there is a great abundance of adequate metrics), but there is a lack of methods to evaluate them (see Sect. 4.5)

In public administration management, the concept of "performance measurement" is far from new, as states Perrin (1998) in his review of performance measurement theory and practice. Performance measures have been widely promoted by governments for more than 30 years, for the purpose of increasing management's focus on achieving results (Winston 1999). This is further demonstrated by the publication in 2001 of "The Performance-Based Management Handbook" by Oak Ridge Institute for Science and Education (ORISE) – U.S. Department of Energy (Performance-Based Management Special Interest Group 2001).

Many authors have tried to address their studies towards the definition of basic rules to assist practitioners in the definition of performance measurement systems (Caplice and Sheffi 1994; Hauser and Katz 1998; Lohman et al. 2004). Nevertheless, a general and organic theory to help in the selection of indicators for the representation of a generic process is still missing. Sect. 1.4.7 synthetically discusses some of the most cited approaches.

The aim of the present chapter is to suggest basic ideas for a general theory of performance indicators.

Process modeling by means of indicators raises many questions: "How many indicators shall we use?", "Is there an optimal set?", "Is this set unique?", "If not, what is the best one (if it exists)?", "Can all these indicators be aggregated in a unique one?", "Are indicators the same as measurements?" etc...

This chapter tries to give an answer to all these questions.

In Sect. 3.2 we provide a definition of the concept of *indicator*. In Sect. 3.3 the condition of *uniqueness* is introduced as well as other basic properties. Practical effects of these properties are explained and discussed by the use of practical examples.

3.2   The formal concept of "indicator"

3.2.1 Definition

The definition of *indicator* is strictly related to the notion of *representation-target*.

A *representation-target* is the operation aimed to make a *context*, or parts of it, "tangible" in order to perform evaluations, make comparisons, formulate predictions, take decisions, etc... Examples of *contexts* are: a manufacturing process (if we are dealing with production management), or a distribution/supply chain (if dealing with logistics), or a market (if dealing with business management), or a result of a competition (if dealing with sports). Given a *context* P, one or more different *representation-targets* Q_P can be defined.

A set of *indicators* S_Q is a tool which operationalizes the concept of *representation-target*, referring to a given *context*:

$$S_Q = \{I_i\}_Q \qquad i = 1, 2, ..., n \qquad n \in \mathbb{N} \tag{3.1}$$

For example, if the *context* is the "logistic process" of a company and the *representation-target* is "the classification of suppliers", the "delivery time" and the "number of defective products" are two of the possible related *indicators*.

In general, it can be shown that, given a *representation-target*, a set of associated *indicators* is not <u>algorithmically</u> generable (Roy and Bouyssou 1993).

The selection of indicators, somehow, creates the problem of the identifying the most significant variables to describe a particular physical phenomenon (Halliday and Resnick 1996).

3.2.2 The representational approach

To better understand the definition of *indicator*, the concept of *measurement* must be reminded. According to the Representation Theory of Measurement, a *measurement* is a "map" from an *empirical relational system* (the "real world") into a *representational relational system* (usually, a *numerical system*) (Roberts 1979; Finkelstein 2003).

Given a set of all possible *manifestations* for a specific characteristic of a well defined representation *context*:

$$A = \{a_1, ..., a_i, ...\} \tag{3.2}$$

and a family of empirical *relations* among the elements of A:

$$R = \{R_1, ..., R_m\} \tag{3.3}$$

then the following *empirical relational system* can be defined:

$$\mathfrak{A} = \langle A, R \rangle \tag{3.4}$$

Analogously, if Z is a set of symbols:

$$Z = \{z_1, ..., z_i, ...\} \tag{3.5}$$

and P is a family of *relations* among the elements of Z:

$$P = \{P_1, ..., P_m\} \tag{3.6}$$

then

$$\Im = \langle Z, P \rangle \tag{3.7}$$

is a symbol relational system.

In general, according to the so called "Symbolic Representation Theory", a measurement is an objective empirical function which maps homomorphically the empirical relational system $\mathfrak{A} = \langle A, R \rangle$ into the symbol relational system $\Im = \langle Z, P \rangle$ (see Fig. 3.1) (Finkelstein 2003).

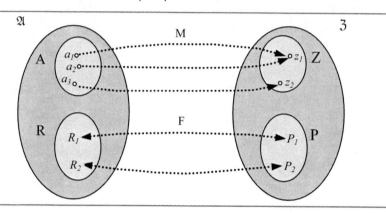

Fig. 3.1. Schematic representation of the concept of *measurement*. A measurement is a function (homomorphism) which maps an *empirical relational system* ($\mathfrak{A}$) into the symbol relational system (typically *numeric*) ($\Im$) (Roberts 1979; Finkelstein 2003). A and R are respectively the set of the *manifestations* and the *relations* in $\mathfrak{A}$. Z and P are respectively the set of the *manifestations* and the *relations* in $\Im$. M and F are the mapping functions from $\mathfrak{A}$ to $\Im$

Two mappings are defined:

$$M : A \to Z \quad \text{(homomorphism)} \tag{3.8}$$

and

$$F : R \to P \quad \text{(isomorphism)} \tag{3.9}$$

so that $M(a_i) = z_i$ is the image in Z of a generic element a_i of A, and $F(R_j) = P_j$ is the image in P of a generic *relation* R_j in R.

M is a homomorphism. The mapping is not one-to-one. Separate but indistinguishable manifestations are mapped into the same symbol.

The "representation code" for $\mathfrak{A}$ is defined as follows:

$$C = \langle \mathfrak{A}, \mathfrak{Z}, M, F \rangle \tag{3.10}$$

The inverse of C is called "interpretation code". z is the symbol of a.

In most applications the mapping is performed into a *numerical relational system*, defined as:

$$\mathfrak{N} = \langle N, P \rangle \tag{3.11}$$

where N is a class of numbers:

$$N = \{n_1, ..., n_i, ...\} \tag{3.12}$$

Usually, N coincides with the set of real numbers $\mathfrak{R}$ and P is a subset of the *relations* on $\mathfrak{R}$.

Referring to the Representation Theory of Measurement, an *indicator* can be considered as a "map" from an *empirical system* (the "real world") into a *symbolic system* (usually, a *numerical system*). However, the mapping between the empirical and symbol *relations* (Eq. 3.9), unlike measurement, is not required:

$$I_Q : a \in A \rightarrow I_Q(a) \in E_Q \tag{3.13}$$

where:
E_Q is the set of elements in the *symbolic system* Z;
A is the set of *manifestations* of the *empirical system*;
a is a *manifestation* of A;
$I_Q(a)$ is the representation of a into the *symbolic system* Z.

Reminding that an *empirical system* is called *relational* if there exists a set of empirical *relations* among empirical *manifestations* (Eq. 3.4), the identification of *relations* is conditioned both by the *context* and the way we model it.

In general, for *indicators*, the mapping of the *empirical system* into a *symbolic* may introduce new *relations* (not present in the empirical system) or modify the existing ones.

In accordance with this approach, three elements have to be considered: the *model* (i.e. the conceptualization of the real world), the *representation-target*, and the rules to determine the related set of *indicators* together with their associated *relations*. The representation does not hold until these three elements are not delineated (see Fig. 3.2).

For example, if we want to identify the winner of a competitive tender:

- the *representation-target* is "finding a winner";

- the *model* is given by "how we evaluate competitors' credentials" (different methods may lead to different classifications);
- the *indicators* and the associated *relations* originate from the rules established for obtaining a final score.

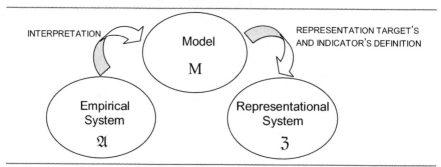

Fig. 3.2. A scheme of the representational approach of an empirical system through the concepts of *model, representation-target,* and *indicators*

On the basis of the representational approach, *measurements* may be interpreted as a subset of *indicators.* The basic difference between *measurements* and *indicators* is the way the relations of the *empirical systems* are mapped. *Indicators* do not require an isomorphism between empirical and symbolic *relations* (Eq. 3.9). This means that, while a *measurement* is certainly an *indicator,* the opposite is not true (see Fig. 3.3).

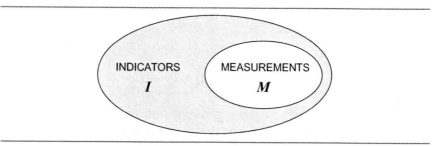

Fig. 3.3. *Measurements* interpreted as a subset of *indicators*

Let us consider, for example, the problem of choosing of a car. The customer preference is an *indicator,* which maps the *empirical system* (different car models) into a *symbolic system* (ranking of the most desired cars). It is not a *measurement.* No order *relation* is defined among *empirical manifestations.* Different subjects usually produce different classifications which, as a consequence, have a different symbolic representation. In this case, there is no *objectivity,* which is typical property of measurements.

In general, if the *symbolic system* is a *numerical system*, an *indicator* is defined as a real value function on the set of *empirical system manifestations*:

$$I_Q : a \in A \rightarrow I_Q(a) \in N \tag{3.14}$$

where N is a class of numbers, defined as in Eq. 3.12.

3.2.3 Basic and derived indicators

An *indicator* is *basic*, if it is obtained as a direct observation of an *empirical system*. Examples of *basic indicators* are the number of defectives in a production line, the number of manufactured parts, the lapse time between events etc...

An *indicator* is *derived* if it is obtained by the synthesis of two or more *indicators*. Examples of *derived indicators* are the total amount of the products created by a set of manufacturing lines, the ratio between defectives and good products for a given time unit in a production line, and so on.

3.3 The condition of "uniqueness"

In general, given a *representation-target*, the related *indicator* (or set of *indicators*) is not univocally defined. The same representation-target can be represented by more independent indicators (or indicators *sets*). This can be shown for both *basic* and *derived indicators* by a series of simple examples.

3.3.1 "Non-uniqueness" for derived indicators

Let us consider an automotive exhaust-systems production plant composed by four equivalent production lines (motorizations): α, β, γ, and δ (Franceschini and Galetto 2004). Fig. 3.4 shows a scheme of the exhaust-system.

In this case, the *context* is the "manufacturing plant" and the *representation-target* is "the identification of the best production line".

Production line performances are defined by the following three *basic indicators*:

- daily production (the number of items produced in a day);

- daily defectiveness (the number of rejected items in a day);
- unavailability equipment ratio (the percentage of breakdown hours in a day).

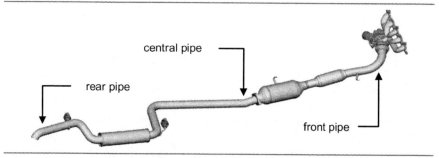

Fig. 3.4. Scheme of an automotive exhaust-system

Given these indicators, at least two different derived indicators which operationalize the assigned representation-target can be found. Let us consider the experimental data reported in Table 3.1.

Table 3.1. Experimental data of four equivalent production lines for exhaust-systems in a manufacturing plant

	Production lines			
Indicators	α	β	γ	δ
daily production [no. per day]	360	362	359	358
daily defectiveness [no. per day]	35	32	36	40
unavailability equipment ratio [%]	4.00%	5.50%	4.50%	5.00%

For each basic indicator we may establish the following rankings:

- *daily production*: $\beta \succ \alpha \succ \gamma \succ \delta$
- *daily defectiveness*: $\beta \succ \alpha \succ \gamma \succ \delta$
- *unavailability equipment ratio*: $\alpha \succ \gamma \succ \delta \succ \beta$

Given these three rankings, the problem is to aggregate them into a unique one. The way to aggregate these three *indicators* is conditioned by a series of constraints, first of all the scale properties and their meaning (Banker et al. 2004).

The assignment of weights, demerits and so on, to reflect the degree importance of each *indicator* is adopted in many circumstances. This is a subjective approach. It suffers from the absence of consistent criteria to determine (a priori) the weighting values. Changing the numerical encoding may determine a change in the obtained results. In this way the analyst of the problem does influence directly the aggregation results. Any conclusions drawn from the analysis on "equivalent" numerical data could be partially or wholly distorted (Roberts 1979; Franceschini and Romano 1999). It is important to remember that the aggregation criteria should be consistent with the data scale properties and their empirical meaning (Roberts 1979; Franceschini et al. 2004). The choice of special codification techniques based on the use of substitution rates or cost utility functions is, in principle, not correct either. The arbitrary application of subjective codification rules can produce radical alterations of final results (Fishburn 1976; Vincke 1992).

Coming back to the problem, we can use two possible derived indicators to determine an overall ranking:

- **Borda's *indicator*** (I_B)

Referring to the order of each basic *indicator* (see Table 3.1), each production line has a rank: 1 for the first position in the ranking, 2 for the second... and n for the last. The Borda score for each line is the sum of every line's rank. The winner is the line with the lowest Borda score.

$$I_B(x) = \sum_{i=1}^{m} I_i(x)$$ (3.15)

where $I_i(x)$ is the ranking obtained by a line x with regard to i-th *basic indicator* and m is the number of indicators used (in this case, $m = 3$).

The winner (the best line x^*) is given by (Borda 1781):

$$I_B(x^*) = \min_{x \in S}\{I_B(x)\}$$ (3.16)

where S is the set of compared lines. In this example, $S \equiv \{\alpha, \beta, \gamma, \delta\}$.

- **Condorcet's *indicator*** (I_C)

For each pair of lines, it is determined how many times a line is higher ranked than an other. Line x is preferred to line y if the number of basic indicators in which x exceeds y is larger than the number of basic indicators in which y exceeds x.

$$I_C(x) = \min_{y \in S-\{x\}}\{i : xPy\}$$ (3.17)

where i is the number of basic indicators in which x exceeds y, and P is the preference operator.

A line that is preferred to all other lines is called the (Condorcet) winner. A (Condorcet) winner is an alternative that, opposed to each of the other alternatives, wins by a majority.

$$I_C(x^*) = \max_{x \in S}\{I_C(x)\} \qquad (3.18)$$

It can be demonstrated that there is never more than one (Condorcet) winner (Condorcet 1785).

Applying Borda's method to data in Table 3.1, we obtain the following results:

$$I_B(\alpha) = 2 + 2 + 1 = 5$$
$$I_B(\beta) = 1 + 1 + 4 = 6$$
$$I_B(\gamma) = 3 + 3 + 2 = 8$$
$$I_B(\delta) = 4 + 4 + 3 = 11$$

According to Eq. 3.16, the final ranking is: $\alpha \succ \beta \succ \gamma \succ \delta$.

The winner (i.e. the line with best overall performance) is line α.

Condorcet's method applied to data in Table 3.1 gives the following results (see Table 3.2):

Table 3.2. Pair-comparisons of data in Table 3.1. Global ranking according to Condorcet's method

	α	β	γ	δ	I_C	Ranking
α	-	1	3	3	1	2°
β	2	-	2	2	2	1°
γ	0	1	-	3	0	3°
δ	0	1	0	-	0	3°

According to Eq. 3.18, the best line is β ($\beta \succ \alpha \succ \gamma \approx \delta$).

The two approaches, though satisfying the same *representation-target*, provide different conclusions about the production plant performances.

A significant aspect regards the use of Borda's *indicator*. It is possible to demonstrate that it is sensitive to "irrelevant alternatives". According to this assertion, if x precedes y in a Borda order, there is no guarantee that x still precedes y if a third alternative z is added (Fishburn 1970; Nurmi 1987).

Consider again an exhaust-system production plant with three production lines $\{\alpha,\beta,\gamma\}$. Suppose they are compared with regard to the daily
production and the daily defectiveness (see Table 3.3).

Table 3.3. Experimental data of three equivalent production lines for exhaust-
systems in a manufacturing plant (1st condition)

	Production Lines		
Indicators	α	β	γ
daily production [no. per day]	367	350	354
daily defectiveness [no. per day]	35	30	37

The two rankings are:

- *daily production*: $\alpha \succ \gamma \succ \beta$
- *daily defectiveness*: $\beta \succ \alpha \succ \gamma$

The resulting Borda scores are:

$$I_B(\alpha) = 1+2 = 3$$
$$I_B(\beta) = 3+1 = 4$$
$$I_B(\gamma) = 2+3 = 5$$

According to Borda's *indicator* (Eq. 3.16), the best line is α.

Now suppose that γ varies its position in the orders (daily production:
from 354 items to 345 items, daily defectiveness: from 37 items to 33
items), while α and β reciprocal position does not change (see Table
3.4).

The new rankings are:

- *daily production*: $\alpha \succ \beta \succ \gamma$
- *daily defectiveness*: $\beta \succ \gamma \succ \alpha$

Table 3.4. Experimental data of three equivalent production lines for exhaust-
systems in a manufacturing plant (2nd condition)

	Production Lines		
Indicators	α	β	γ
daily production [no. per day]	367	350	345
daily defectiveness [no. per day]	35	30	33

The new resulting Borda scores are:

$$I_B(\alpha) = 1 + 3 = 4$$
$$I_B(\beta) = 2 + 1 = 3$$
$$I_B(\gamma) = 3 + 2 = 5$$

In this case, the best line is β, even if the performances of α and β do not change. This demonstrates the Borda's indicator sensitivity to "irrelevant alternatives" (line γ). In fact, in both conditions. The third alternative plays a marginal role.

On the other hand, it can be shown that the Condorcet's method does not guarantee the property of "transitivity" between *relations* (Fishburn 1977).

Consider, for example, the exhaust-system production plant with three lines $\{\alpha, \beta, \gamma\}$. Suppose they are compared with regard to the daily production, the daily defectiveness, and the unavailability equipment ratio (see Table 3.5).

Table 3.5. Experimental data of three equivalent production lines for exhaust-systems in a manufacturing plant

Indicators	Production Lines		
	α	β	γ
daily production [no. per day]	365	362	359
daily defectiveness [no. per day]	35	32	34
unavailability equipment ratio [%]	5.50%	6.00%	4.50%

The resulting rankings are:

- *daily production*: $\alpha \succ \beta \succ \gamma$
- *daily defectiveness*: $\beta \succ \gamma \succ \alpha$
- *unavailability equipment ratio*: $\gamma \succ \alpha \succ \beta$

Condorcet's method gives the results in Table 3.6.

In this case, there is no winner. The transitivity property is not satisfied. According to direct comparisons, it results that:

$$\alpha \succ \beta; \qquad \beta \succ \gamma; \qquad \gamma \succ \alpha$$

In conclusion, I_B and I_C are independent *indicators*. There is no mathematical transformation which maps Borda scores into Condorcet scores, or vice versa, maintaining the global order (see Fig. 3.5) (Fishburn, 1970, 1977).

Table 3.6. Pair-comparisons of data in Table 3.5. Global ranking according to Condorcet's method

	α	β	γ	I_C	Ranking
α	-	2	1	1	$1°$
β	1	-	2	1	$1°$
γ	2	1	-	1	$1°$

The results obtained by alternatively applying Borda's and Condorcet's *indicators* are different, although they have been provided for the same *representation-target*. We can deduce that the representation condition is valid for more than one *indicator*. In general, *uniqueness* is not guaranteed for *derived indicators*.

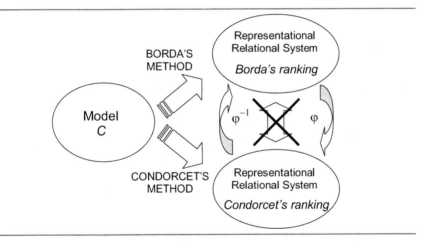

Fig. 3.5. Schematic representation of the "independence" between Borda's indicator and Condorcet's indicator

3.3.2 "Non-uniqueness" for basic indicators

Let us consider again the four production lines $\{\alpha, \beta, \gamma, \delta\}$ introduced before. Suppose that the comparison is performed with regard to daily defectiveness. The representation-target is "*identifying the line with the lower defectiveness*". At least, two different indicators can be adopted:

- The indicator "*number of rejected parts*" (I_R)

For each line, I_R is given by the number of rejected parts within the daily production. The best line ($x*$) is the one with the minimum value:

$$I_R(x*) = \min_{x \in A}\{I_R(x)\} \tag{3.19}$$

being $A \equiv \{\alpha, \beta, \gamma, \delta\}$.

- The indicator "*number of defects detected*" (I_D)

For each line, I_D is given by the number of defects detected during a full control of the daily production. Incidentally, a defect is a failed fulfilment of one specification. A defective element may include one or more defects, and not all types of defects (especially if they are minor) lead to a part reject. Getting back to the point, the better line is the one with the minimum value I_D:

$$I_D(x*) = \min_{x \in A}\{I_D(x)\} \tag{3.20}$$

Here again, there is no mathematical transformation which biunivocally links the two indicators. So, the same representation target can be represented by more than one basic indicator. In general, *uniqueness* is not guaranteed for *basic indicators* as well.

By way of example, let us consider the daily defectiveness data (defects and rejected parts) of the four production lines (Table 3.7).

Table 3.7. Daily defectiveness indicators of the four production lines $\{\alpha, \beta, \gamma, \delta\}$

	I_S	I_D
α	35	43
β	25	39
γ	17	45
δ	21	25

Considering the indicator I_R, the best line is γ:

$$I_R(x*) = \min_{x \in A}\{35, 25, 17, 21\} = 17 \tag{3.21}$$

On the other hand, considering the indicator I_D, the best line is δ:

$$I_D(x*) = \min_{x \in A}\{43, 39, 45, 25\} = 25 \tag{3.22}$$

Changing the reference indicator, the result of the evaluation changes.

The use of the indicator I_R, rather than I_D can have different consequences on the behaviour of the production line administrator. I_R focuses on minimising the number of rejected parts (for instance, tolerating minor defects); while I_D focuses on optimizing the process phases, to reduce all sorts of defects. Even if they represent the same representation-target, different indicators will affect different actions and decisions. Sect. 5.9 will analyse the problem of the indicators impact onto processes.

3.3.3 Remarks about the condition of "uniqueness"

The non fulfilment of uniqueness condition implies a series of consequences in the use of indicators. The most evident is that there is an arbitrary choice in setting up the mapping into the symbolic system. This entails that, given two or more indicators for a specified representation-target, there may be no transformation from one indicator into another. This causes, for example, analogous representation-targets which might not be comparable, if represented by different indicators.

On the other hand, it is interesting to remember that also the *measurements* do not fulfil the condition of uniqueness (Roberts 1979). For *measurements*, the requirement of homomorphism for mapping empirical *manifestations* and isomorphism for mapping *relations* (Eqs. 3.8 and 3.9) defines a *"class of equivalent scales"*. Each equivalent scale can be mapped into another. All the possible transformations form the so called *"class of admissible transformations"* (Finkelstein 2003).

The unclear objectification of the *model* and/or the incomplete definition of the *representation-target*, as well as the non fulfilment of the condition of *uniqueness*, are at the basis of the concept of *uncertainty*. In particular, *uncertainties* of *measurement* are considered to be imperfections of the measurement process and/or the result of incorrect determinations of empirical observations or empirical laws (Finkelstein 2003). A similar concept can be defined for *indicators*.

3.3.4 Condition of "uniqueness" by specializing the representation-target

It is easy to show that a deeper specialization of the representation-target does not imply the automatic removal of the condition of *non-uniqueness*.

Let us consider again the four lines $\{\alpha,\beta,\gamma,\delta\}$ introduced in the example reported in Table 3.1. We showed that "the identification of the line with the lower defectiveness" yields at least two different indicators.

Now, let us try to further specialize the representation-target definition in order to eliminate the *non-uniqueness* of related *indicators*. A more specialized definition may be "*identifying the line with the lower number of rejected parts*". Also in this case, at least two different *indicators* can be adopted:

- The indicator "*number of rejected parts which cannot be reworked*" (I_{Snr})

For each line, I_{Rnr} is given by the daily number of rejected parts which cannot be reworked. The best line ($x*$) is the one with the minimum value:

$$I_{Snr}(x*) = \min_{x \in A}\{I_{Snr}(x)\} \qquad (3.23)$$

where $A \equiv \{\alpha,\beta,\gamma,\delta\}$

- The indicator "*number of rejected parts with two or more defects*" (I_{Sdd})

For each line, I_{Sdd} is given by the daily number of rejected parts with two or more detected defects. The best line ($x*$) is the one with the minimum value:

$$I_{Sdd}(x*) = \min_{x \in A}\{I_{Sdd}(x)\} \qquad (3.24)$$

where $A \equiv \{\alpha,\beta,\gamma,\delta\}$.

In general, we can affirm that a semantic specialization of the representation-target implies a more accurate definition of the related indicator (or set of indicators), never reaching the condition of *uniqueness*. In fact, we can try to further specialize the *representation-target*: "identifying the line with the lower average number of rejected parts, with two or more defects". This new definition does not imply the *uniqueness* of the indicator. We can still define at least two new different indicators. For example, we can adopt an indicator which excludes from the rejected elements those which only are originated by a breakdown, and another which includes them. And so on.

The result is that an univocal definition of an indicator can never be obtained. The remaining (small) differences, after the *representation-target* specialization, will contribute to *uncertainty*.

In conclusion, the condition of uniqueness can be considered from two possible points of view. The first one concerning *definition* and the second *representation*. The semantic specialization of the *representation-target* concerns only the first aspect.

3.3.5 The choice of the best set of indicators

An immediate consequence of the *non-uniqueness* condition is that a representation-target can be described by different set of indicators. This leads to the need of establishing a series of rules (or empirical procedures) to individuate the set of indicators which better embodies a given representation-target. The choice of the best set of indicators involves the analysis of the possible *impact* that indicators will produce on the system. A different set of indicators may differently influence the overall behaviour of a system with uncontrollable consequences (Barnetson and Cutright 2000; Hauser and Katz 1998).

To select indicators, two different typologies of properties should be considered:

- *basic properties*, directly related to the mathematical definition of *indicator* (uniqueness, exhaustiveness, monotony, and non-redundancy, etc...) (Roberts 1979; Roy and Bouyssou 1993);
- *operational properties*, related to their application practice (validity, robustness, usefulness, integration, economy, compatibility, etc...) (Caplice and Sheffi 1994).

According to each application case, the final choice must be addressed towards the set which better meets the two families of properties and generates the most "effective" *impact*. Chap. 4 presents a taxonomy of the indicator properties.

4. Performance indicators properties

4.1 Introduction

The previous chapters described indicators as helpful tools in collecting information and analysing the evolution of complex systems/processes.

Generally, a "set of indicators", according to the definition given in Chap. 1, has to represent all the different process aspects. It is a "representation model" to support evaluations and decisions on the process itself. Each model is a *"representation scheme of many phenomena, extracted from their context to support the analysis"*[4] (Roy and Bouyssou 1993).

Selection of good indicators is not an easy activity. The success usually depends on the experience and the initiative of the people performing it. Actually, there are not general and organic methods to help defining indicators. Chap. 5 will discuss some techniques. Furthermore, very often indicators are adopted without a preliminary analysis of the impact they can produce.

This chapter provides a categorization of the indicators peculiarities, presenting a taxonomy which identifies the properties that indicators should have, to represent a generic system/process properly.

4.2 Local and aggregated performances

Indicators should be defined considering two levels of detail:

- *Single (local) level.* The indicator represents a particular aspect of the system. Each indicator is a link between the "empirical manifestations"

---

[4] This definition is extended to all the categories of models. For instance the physical models, the analytical simplified models, the dynamic models, etc.

of the examined aspect, and some corresponding "symbolic manifestations"[5].

- *Aggregated (global) level.* The whole system can be represented by a set of indicators. The set should represent the most important system aspects, without forgetting, omitting or misunderstanding them, and with no redundancies[6].

Correspondingly, for a system modelled by indicators two kinds of performances can be defined:

- *Single (local) performance.* It is the system performance, from the local point of view of a single indicator (representing a single system aspect);
- *Aggregated (global) performance.* It is the performance of the whole system, or a specific macro-portion. The whole system performance is based on the aggregation of local performances. For complex systems, it is not always reasonable to summarize the global performance in single information, putting together local performances by algorithmic relations. Frequently, relations of that sort can be questionable since they are based on "dangerous" simplifications. To better explain the concept, we provide an example.

Example 4.1 In a university entrance exam, the best 200 students are selected according to the following three indicators:

I_1: mark of the high school leaving qualification;
I_2: result of a written entrance test;
I_3: result of an oral entrance test.

Each of these indicators is expressed by a numerical value between 0 and 100. The total/aggregated performance is calculated considering a weighted average value. The three indicators (I_1, I_2, I_3) weights are respectively 5/10, 3/10 and 2/10. Consequently, the total score (I_{TOT}) is given by Eq. 4.1:

$$I_{TOT} = \frac{5}{10}I_1 + \frac{3}{10}I_2 + \frac{2}{10}I_3 \qquad (4.1)$$

The high school leaving qualification mark (I_1) has the largest weight, and – on the contrary – the result of the oral entrance test (I_2) has the lowest. In this case, the choice of the weights is questionable, but has important effects on the aggregated performance (I_{TOT}). Different weights define different selection strategies.

---

[5] Expressions such as empirical manifestation or symbolic manifestation refer to the Representation Theory of Measurement, explained in Chap. 3.

[6] This statement implicitly refers to the concept of counter-productivity, formalised in Sect. 4.6.1.

4.3 General remarks

We remind that, according to the *Representation Theory of Measurement*, an indicator maps the manifestations of an empirical system onto corresponding manifestations of a symbolic system (see Fig. 4.1). The empirical system definition is tightly related to the concept of *representation-target* (see Sect. 3.2.1).

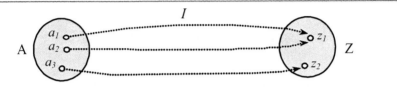

Fig. 4.1. Schematic representation of the concept of indicator. An indicator (*I*) is defined as a homomorphism from a set of real empirical manifestations (A) into a set of symbolic manifestations (Z). In formal terms *I*: A→Z

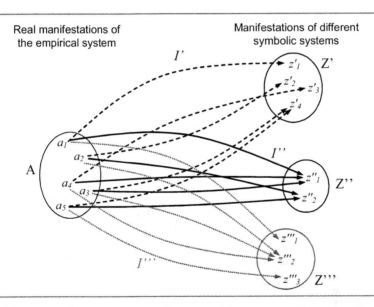

Fig. 4.2. Schematic representation of the condition of *non-uniqueness*. The same representation target is operationalized by three different indicators (*I'*, *I''* and *I'''*). Some empirical manifestations, indistinguishable according to one indicator, can be distinguished by another one (for example the manifestations a_3 and a_5 are undistinguished by *I'*, but distinguished by *I''* and *I'''*)

We can remark that for homomorphisms the mapping is not one-to-one. Manifestations which are indistinguishable, according to the representation-target, are mapped into the same symbol (in Fig. 4.1 manifestations a_1 and a_2 are considered indistinguishable and therefore they are mapped into the same symbol z_1).

Chap. 3 has extensively discussed the condition of *non-uniqueness* of the representation. The scheme in Fig. 4.2 summarizes the concept.

Example 4.2 In a manufacturing company producing hydraulic valves, the purpose is *"to improve the quality of the produced goods"*. The following indicator set is implemented to operationalize this representation target:

I_1 *number of units produced;*
I_2 *(monthly) number of units categorized as defective, and rejected.*

A second possible indicators set is given by:

I'_1 *number of units produced by the first of 4 production lines;*
I'_2 *average percentage of detective units: result of a spot check on the 5% of total production.*

Since different indicators sets may refer to the same representation-target, some questions arise: "what is the best way of selecting them?"; "when is the representation exhaustive?".

A tool that algorithmically generates a set of indicators, for a specific representation-target, is not actually conceivable (Roy and Bouyssou 1993). Such a tool would have to implement a chain of operations normally carried on by the system "modeller": (1) definition of a set of indicators; (2) preliminary test; (3) usability verification; (4) correction of the model; (5) further verification.

In conclusion, even if there are many ways for representing the same process, the "best" one cannot be *a priori* identified. This chapter will describe some properties and rules to support the selection and aggregation of indicators and the verification of the representation model.

As shown in Sect. 3.2.2 there is a strong link between the concept of measurement and indicator. On the basis of the Representation Theory, measurements can be interpreted as a subset of indicators. The following examples emphasize this aspect.

Example 4.3 The wheelbase of a motor vehicle (the geometrical distance between two car-axles) is a dimensional *measurement*, and therefore also an *indicator*. The relationships among symbolic manifestations (numerical distance values) are isomorphically linked to the relations among physical manifestations (physical distance).

Example 4.4 The representation-target *classification of objects depending on their volume*, is operationalized using the indicator *object volume* (expressed in cm^3), which is also a *measurement*. The relationships of equivalence (volume A is different from volume B), order (volume A is lager than volume B), and ratio (volume A is "*x*" times larger than volume B), included in the empirical relational system, are equally included in the symbolic relational system (Finkelstein 2003; Franceschini 2001).

Example 4.5 The indicator *comfort of a car seat*, codified by a qualitative numeric scale from 1 to 5, is not a ratio *measurement*. Ratio relationships (like: seat A is two times more comfortable than seat B) among symbolic representations are not necessarily kept in the real world.

Example 4.6 Let us consider the representation-target *classification of the students of a class*, operationalized by the indicator *name of the student*. This indicator associates each student (empirical manifestation) to the corresponding name (symbolic manifestation). In nature there is no order relation among the empirical manifestations (the students), which corresponds to the alphabetical order relation among the symbolic manifestations (names). So, the *name of the student* is only an indicator, not a measurement. The order relation among symbolic manifestations does not correspond to any existing relation among real manifestations.

4.4 Indicators classification

In general, any system can be represented using indicators. Usually, the more complex the system, the larger the number of selected indicators, and their variety (basic, derived, objective, subjective, etc...). Even if the information available is often a lot, the number of indicators should not be too large, in order not to complicate the system representation (Melnyk et al. 2004).

The following section discusses the classification presented in Chap. 3, supplying it with practical examples. Indicators are classified in *objective, subjective, basic* and *derived*.

4.4.1 Objective and subjective indicators

Indicators can be classified in two main categories: *objective* and *subjective*.

Objective indicators. They objectively link empirical manifestations to symbolic manifestations. The mapping does not depend on the subject who performs it.

Example 4.7 Let us consider the indicator: *quantity of goods produced in a plant*. The empirical manifestation (production) can objectively be connected to a corresponding symbolic manifestation (number of goods produced). If there is no counting mistake, different people (or automatic devices even) determine the same value.

Objective indicators are not necessarily measurements. Example 4.6 shows this aspect. Name assignment concerns the concept of *conventional objectivity*, discussed in Sect. 5.9.

Subjective indicators. Empirical manifestations are subjectively mapped into symbolic manifestations, depending on subjective perceptions or personal opinions. Therefore, different people can map the same empirical manifestation into different symbolic manifestations. Let us consider, for example, the following indicators: the *individual vote at the elections*, and the *evaluation of the design of a car*. Such indicators are usually confined to personal perceptions and opinions.

Example 4.8 The representation-target *evaluation of the design of a car* can be operationalized by the indicator *quality of the design*, codified with a 5 level scale (1-*very bad*; 2-*poor*; 3-*fair*; 4-*good*; 5-*excellent*). The indicator is *subjective* because the same empirical manifestation (a particular car design) can be associated to different symbolic manifestations (the 5 scale levels), depending on the subject.

Since subjective indicators provide essential information about the individuals' behaviour and perceptions, they are often used and studied by many disciplines in the area of *Social*, *Behavioural* and *Cognitive Sciences* (Nunally 1994). The numeric encoding is a common way to make the information practical for the user. However – when the relations among symbolic manifestations do not correspond to relations among empirical – this sort of conversion may distort the analysis results (Roberts 1979; Franceschini and Rossetto 1995; Franceschini and Romano 1999). Sometimes, the symbolic relations are wrongly "promoted", with reference to the empirical ones, because of the introduction of additional hypothesis in the mapping (as seen in Example 4.6). So, all the hypotheses made must be carefully verified (Narayana 1977; Franceschini 2001). If the relations among the empirical manifestations do not correspond to the relations among symbolic manifestations, we are dealing with *indicators* – rather than *measurements*.

4.4.2 Basic and derived indicators

As explained before (Sect. 3.2.3), indicators can also be subdivided in two more categories:
* *Basic indicators*. They are obtained from a direct observation of an empirical system.

- *Derived indicators.* They are obtained combining the information of one or more "source" indicators (basic or derived).

Example 4.9   Let consider the derived indicator: percentage of defectives in a production line, given by:

$$I_3 = \frac{I_1}{I_2}$$ ⟵   (number of defective units)
         ⟵   (total number of produced units)   (4.2)

Example 4.10 An organization for the environmental protection asks two local Agencies - A and B - to estimate the pollution level of the exhaust emissions of a motor vehicle, on the basis of four pollutants concentrations:

I_{NO_X} : the concentration of NO_X in the exhaust emissions [$\mu g/m^3$];

I_{HC} : the concentration of un-burnt hydrocarbons [$\mu g/m^3$];

I_{CO} : the concentration of CO [$\mu g/m^3$];

$I_{PM_{10}}$: the concentration of (PM_{10}) [$\mu g/m^3$].

Agency A maps each concentration into a 3 level scale (1-harmless; 2-acceptable; 3-unacceptable for the human health), and specifies 4 corresponding derived indicators ($I'_{NO_X}, I'_{HC}, I'_{CO}, I'_{PM_{10}}$). Then it defines an additional derived indicator, I^A_{TOT} , assuming the maximum value of them (see Chap. 2).

$$I^A_{TOT} = \max \left\{ I'_{NO_X}, I'_{HC}, I'_{CO}, I'_{PM_{10}} \right\}$$ (4.3)

Fig. 4.3 shows the aggregation principle.

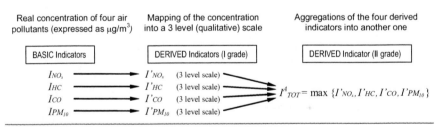

Fig. 4.3. *Basic* and *derived* indicators implemented by Agency A.

Agency B maps the concentration of each pollutant into a 5 level scale, and defines 4 corresponding derived indicators (I''_{NO_X} , I''_{HC} , I''_{CO} , $I''_{PM_{10}}$). Then it specifies an additional derived indicator aggregating the previous ones by the average value of them:

$$I_{TOT}^{B} = \frac{\left(I_{NO_X}^{"} + I_{HC}^{"} + I_{CO}^{"} + I_{PM_{10}}^{"}\right)}{4} \qquad (4.4)$$

Agency A estimates the pollution level on the basis of concentration "peaks", while Agency B on the basis of their combination (a sort of the "principle of superimposition of the effect").

The example shows that, given a representation-target, the same *basic* indicators can be aggregated in dissimilar ways. For a more detailed description of the most common aggregation techniques for air quality indicators, see Sect. 2.3. Additionally, the example shows that *derived* indicators may be aggregated (again) in a *derived* indicator of higher grade (see Fig. 4.4). By extending this concept to the limit, we can imagine to define a "super-indicator", synthesising all the aspects of the system investigated. After defining single *basic* indicators, the real challenge is to group them together, in order to set up a model providing general information on the system global performance (Melnyk et al. 2004).

Chap. 5 exhaustively discusses whether this recursive synthesis process is always reasonable.

In conclusion, it is interesting to note that indicator I_{TOT}^{B} introduces a new relation among the symbolic manifestations (interval properties), which empirically does not exist. As they are defined, indicators scales support only the order relation (Franceschini 2001).

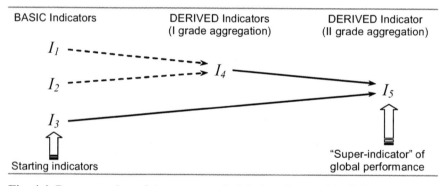

Fig. 4.4. Representation of the concept of *global performance*. All the starting *basic* indicators are aggregated into a (global) derived indicator

4.4.3 The representational approach for derived indicators

The concept of *derived indicator* can also be interpreted according to the Representation Theory. The empirical system of a derived indicator is given by the combination of the "source" indicators symbolic manifestations. This combination is then homomorphically mapped into further symbolic manifestations (see Fig. 4.5).

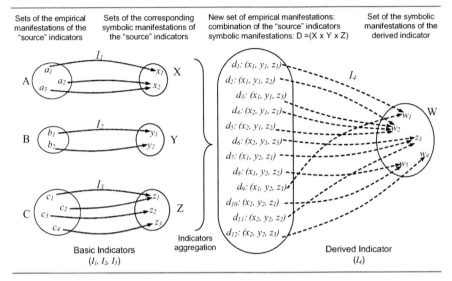

Fig. 4.5. Schematic representation of a derived indicator, according to the Representation Theory

The aggregation of several indicators into a derived indicator is not always easy to achieve, especially when the information to synthesise is assorted. Let us consider the following examples.

Example 4.11 To investigate the *general condition of a manufactory company*, different indicators representing various aspects of the system – as *profit margin, throughput, market share, customer satisfaction* etc... – are implemented. The aggregation of these heterogeneous indicators can be very complex and based on questionable simplifications.

Example 4.12 A manager of a manufacturing company decides to reengineer a particular product, in order to improve the quality and to reduce possible failures. First he searches for the most significant product defects, and then tries to sort them in a priority order. To get that purpose, each possible failure is associated with two indicators: the index of the estimated gravity (g), and the frequency (f).

These indicators are aggregated into a single derived indicator called priority index (*PI*):

$$PI = f * g \qquad (4.5)$$

where:

f failure frequency ($0 \le f \le 1$) (basic indicator)
g estimated gravity index (basic indicator)

Detected failures are sorted by the value of *PI*, in descending order. This method looks easy, but it can guide to divergent conclusions. For instance, the order of priority may change unexpectedly, depending on the encoding of the estimated gravity index *g*. Let us consider the following example: the gravity index is encoded according to two scales (scale A and scale B), each of them is divided in four ordered levels (L_1, L_2, L_3, L_4; see Fig. 4.6).

The derived indicator *PI* is determined in two conditions (Failure n.1 and Failure n.2, in Fig. 4.6), evaluating *g* using either the previous scales. As shown, the ordering of two generic defects changes depending on the *g* scale!?!

Index of gravity (*g*)	L_1	L_2	L_3	L_4
SCALE A	1	2	3	4
SCALE B	4	5	6	7

	f	*g*			*PI = f*g*	
		Level	Scale A	Scale B	Scale A	Scale B
Failure n.1	0.5	L_2	⇒ 2	⇒ 5	0.5*2= 1	0.5*5= 2.5
Failure n.2	0.3	L_4	⇒ 4	⇒ 7	0.3*4= 1.2	0.3*7= 2.1

Fig. 4.6. Product reengineering. Priority index for different types of failure. The ordering of two failures changes depending on the scale of the estimated gravity index (*g*)

The reason of this paradox is that by mapping the estimated gravity index (empirical manifestation) into a number (symbolic manifestation) we introduce a new relation (ratio relation), which empirically does not exist (order relation only: $L_4 > L_3 > L_2 > L_1$). According to the ratio relation, for example, level L_4 is equivalent to four times level L_1 (Finkelstein 2003; Franceschini 2001). Since the ratio property among symbolic manifestations does not correspond to any existing relation among real manifestations, *g* is an indicator, not a measurement. In these cases, different mappings of the same empirical manifestations may lead to paradoxical situations, like the one described.

In conclusion we can remark that it is not uncommon to aggregate indicators with different properties of scale (like *f* and *g*), which lead to undesired consequences.

4.5 A brief outline of the indicators properties in the literature

We analyzed the existing literature with the purpose of finding the major properties that indicators should satisfy for a suitable process representation. We found lots of indicators properties and definitions often unstructured, and presented in different ways by different authors. Properties are often described without a formal mathematical approach. For manufacturing, as well as for other business functions in general, researchers have identified several criteria to consider when selecting individual performance indicators.

Table 4.1. Comparison of different individual properties of indicators (Caplice and Sheffi 1994). With permission

Requirements of the indicators	Properties defined by different authors					
	Mock and Groves (1979)	*Edwards (1986)*	*Juran (1988)*	*NEVEM (1989) AT Kearney (1991)*	*Mentzer and Konrad (1991)*	*Caplice and Sheffi (1994)*
Does the indicator capture actual events and activities accurately?	Valid	Reliable		Valid		Validity
Does the indicator control for inherent errors in data collection? Is it repeatable?	Reliable				Measurement Error	Robustness
Is the indicator using a correct type of numerical scale for the values?	Scale Type					Behaviourally Sound
Is the indicator still accurate if the scale is transformed to another type?	Meaningful					Behaviourally Sound
Do the benefits outweigh the costs of using the indicator?	Economical Worth	Cost/ Benefit	Economical	Profitability		Economy
Will the indicator create incentives for improper or counterintuitive acts?	Behavioural Implications				Human Behaviour	Behaviourally Sound
Does the indicator use data currently available from the existing ones?		Available				Compatibility
Is the indicator compatible with the existing information system and flow?			Compatible to existing systems	Compatible		Compatibility
Does the indicator provide a guide for an action to be taken?		Useful		Utility		Usefulness
Can the indicator be compared across time, location, and organizations?		Consistent	Apply Broadly	Comparable	Comparable	Robustness
Is the indicator viewed and interpreted similarly by all affected parties?			Uniform Interpretation & agreed upon basis			Robustness
Is the indicator simple and straightforward enough to be easily understood by those affected?			Understandable			Usefulness
Does the indicator include and measure all of the important components and aspects of the system?				Covering Potential	Under- determination	Integration
Does the indicator promote coordination across the various players in the supply chain?						Integration
Is the indicator of a sufficient level of detail and precision for a decision maker?				Accurate		Level of detail

Table 4.1 reports the classification suggested by Caplice and Sheffi (Caplice and Sheffi 1994). The table summarizes the major properties of indicators, presented in literature by various authors. Properties are presented without making any distinction among basic, derived or sets of indicators.

4.6 A proposal of a taxonomy for indicators properties

In this section, we present a taxonomy of indicators' properties, based on the classification described in Sect. 4.4 and summarized in Table 4.2. Properties are classified into four groups: *general* properties, properties of *derived* indicators, properties of *sets* of indicators, *accessory* properties. These properties can represent a useful tool to select and evaluate performance indicators in different contexts.

Table 4.2. Indicators properties taxonomy, based on the classification described in Sect. 4.4

Category	Properties	Short description
General properties	Consistency with the representation-target	The indicator should properly represent the representation-target.
	Level of detail	The indicator should not provide more than the required information.
	Non counter-productivity	Indicators should not create incentives for counter-productive acts.
	Economic impact	Each indicator should be defined considering the expenses to collect the information needed.
	Simplicity of use	The indicator should be easy to understand and use.
Properties of sets of indicators $S = \{I_i, I_j, I_k\}$	Exhaustiveness	Indicators should properly represent all the system dimensions, without omissions.
	Non redundancy	Indicators set should not include redundant indicators.
Properties of derived indicators $(I_i, I_j, I_k) \Rightarrow I_{TOT}$	Monotony	The increase/decrease of one of the aggregated indicators should be associated to a corresponding increase/decrease of the derived indicator.
	Compensation	Changes of different aggregated indicators may compensate each other, without making the derived indicator change.
Accessory properties	Long term goals	Indicators should encourage the achievement of process long-term goals.
	Impact on the stakeholders	For each indicator the impact on process stakeholders should be carefully analysed.

It can be reasonably assumed that a large part of the properties found in the literature (see Table 4.1), can be incorporated in this scheme of classification.

4.6.1 General properties

The following properties refer to single indicators. They are effective both for basic and derived indicators.

Consistency with the representation-target

According to the general definition given in Sect. 3.2.2, every indicator should properly operationalize a representation-target. This mapping should be thoroughly verified before using the indicator (Denton 2005).

This concept is expressed well in the Example 4.13.

Example 4.13 Referring to the representation-target *sales of a manufacturing company*, the indicator I_v – *total number of goods sold in the* <u>*whole year*</u> – is used to represent the process. Later, company managers realize that <u>quarterly</u> information on sales would be more useful for estimating the seasonal trend. Consequently, a new indicator I'_v representing the total number of quarterly sold goods replaces the first one.

According to the representation-target, the second indicator is more accurate than the first one. It comprehends some important empirical manifestations (quarterly information on sales), ignored by I_v.

Selecting indicators is not an easy task because they should represent all the process dimensions, without omissions or redundancies. Otherwise, the model is inaccurate or incomplete and may not fulfil some properties, like "exhaustiveness" and "non-redundancy".

Level of detail (resolution)

An indicator with excessive level of detail provides more than the achievable (or required) information, so it could complicate the analysis and could be economically wasteful. Even more, if an indicator maps two empirical manifestations – not distinguished according to a representation-target – into different symbolic manifestations, then the *level of detail is excessive* (see Fig. 4.7-a). To realize whether the indicator mapping resolution level is higher than necessary, we have to carefully analyse the representation-target definition level.

In formal terms:

<u>If</u> $I_k(a) = z_1;\ I_k(b) = z_2$, being $z_1 \neq z_2$

<u>and if</u>   the empirical manifestations of the states "*a*" and "*b*" are <u>not</u> <u>distinguishable</u>, according to the representation-target

<u>then</u> I_k has an excessive level of detail (resolution)

being:

a and *b*   two states of the process;

z_1 and z_2 corresponding symbolic manifestations.

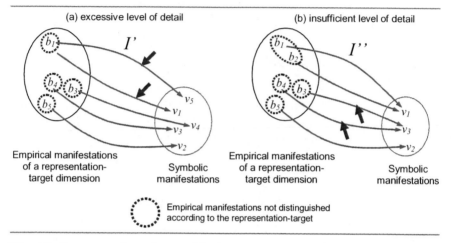

Fig. 4.7. Representation scheme of indicators with excessive (a) and insufficient (b) level of detail

Example 4.14 A manufacturing company produces metal screws. The representation-target is *the production rate of the company*. The indicator *I* represents the "daily weight of produced screws". If the indicator's accuracy is ± 1 g/day – when ± 10 kg/day would be enough – then the level of detail is excessive.

In other words, if the indicator mapping is more accurate than required, two different empirical manifestations, indifferent according to the representation-target, can be unreasonably distinguished (i.e. two different screws daily productions: *I*(*a*)=653.321 kg/day and *I*(*b*)=650.023 kg/day).

Example 4.15 Let us consider the indicator described in Fig. 4.8, to operationalize the representation-target *external design of a car*. The scale categories number (15) can be excessive, considering the normal subjects discrimination ability: let us think of the very small aesthetic differences between two empirical manifestations mapped into two consecutive categories (symbolic manifestations). In other words, the indicator's resolution is excessive.

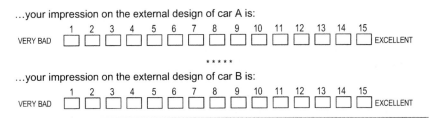

Fig. 4.8. Indicators with an excessive level of detail. The connection between empirical manifestations (impression on the external car design) and symbolic manifestations (scale categories) can be difficult, due to the excessive number of categories

On the other hand, an indicator resolution could be lower than required. In such a situation important information on the process investigated could be lost. Even more, if an indicator maps two empirical manifestations, which should be distinguished according to the representation-target, into the same symbolic manifestation, then the level of detail could be insufficient (see Fig. 4.7-b).

In formal terms:

If $I_k(a) = z_1$; $I_k(b) = z_2$, being $z_1 \approx z_2$

and if the empirical manifestations of the states "a" and "b" are distinguishable, according to the representation-target

then I_k has an insufficient level of detail

being:
a and b two states of the process;
z_1 and z_2 corresponding symbolic manifestations.

Non counter-productivity

Before introducing the concept of *non counter-productivity*, we should make some preliminary remarks. Typically, in a company or in a process managed by indicators, managers and employees focus their attention on indicators linked to short-term rewards or bonuses[7], overlooking the global targets of their tasks. This behaviour can sometimes be counter-productive for the achievement of long-term goals. Even more, indicators may differently *impact* the overall behaviour of a system with uncontrollable consequences.

---

[7] a bonus is an internal incentive, given by the company organization to managers and employees who concur to increase some specific indicators.

Example 4.16 The main purpose of a construction company is to reduce the construction work time, in order to take a competitive advantage. This purpose may generate some counterproductive actions:

- to save time, employees do not obey safety rules (i.e. they do not use the safety helmets and harness);
- working vehicles, rushing around the building site to save time, become dangerous for the public safety;
- customer satisfaction decreases, because the result of the work is poor, due to the excessive speed up.

In this case, focusing too much on a single *dimension* of the process can be counter-productive in general terms.

The idea of counter-productivity can be shown as follows. Some indicators (I_h, I_i, I_m) are aggregated in a derived indicator (I_{TOT}), representing the *global performance*. If the increase of a specific source indicator (I_k) is associated with the decrease of one or more indicators (for example I_i, I_l, I_m), determining a decrease of the global performance (I_{TOT}) too, then I_k is counter-productive. This definition entails that the symbolic manifestations of the source indicators are defined at least on an ordinal scale. Which means that the scale used to measure the source indicator allows local comparisons among symbolic manifestations, like $I(a') > I(a)$. The concept of counter-productivity is meaningless for indicators represented in scales without order relation (for example category scales: Yes-No, A-B-C, etc...). Fig. 4.9 provides a representation scheme of the concept of counter-productivity.

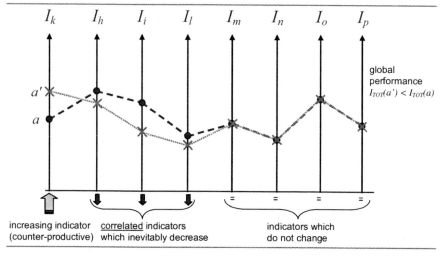

Fig. 4.9. Concept of counter-productive indicator

To assess counter-productivity, process indicators must be well known by users. In other terms, indicators and associated bonuses must be familiar to managers and employees involved in the process. If counter-productive indicators are linked to bonuses, and are simpler to be increased than others, the attention of the employees may dangerously focus on them.

The concept of counter-productivity may be defined in more formal terms. Let suppose that a process represented by n indicators aggregated into a derived indicator I_{TOT}, is in the state "a". If the process skips from state "a" to state "b", and the increase of a source indicator (I_k) is correlated to the decrease of one or more other source indicators $(I_h, I_i, I_m..)$:

$$I_k(b) > I_k(a)^{[8]}$$
$$I_h(b) < I_h(a)$$
$$I_i(b) < I_i(a)$$
$$I_m(b) < I_m(a)$$
$$\ldots \qquad \ldots$$

so that the global performance of the derived indicator I_{TOT} decreases – $I_{TOT}(b) < I_{TOT}(a)$ – then I_k is counter-productive.

When testing the counter-productivity property, the most difficult aspect is to identify the conceptual or empirical correlation between indicators involved.

Example 4.17 To estimate the costumer satisfaction, a call-center uses several indicators. Two of them are the following:

I_1 *average number of rings before answering the phone*;
I_2 *percentage of unanswered calls.*

These two indicators can be counter-productive because employees can "game" the process answering the phone immediately and then putting the call on hold before starting the conversation.

Although that behaviour increases the value of selected indicators, it is absolutely counter-productive according to other indicators of customer satisfaction. For example, the *number of exhaustive answers*, the *courtesy*, the *number of queued calls* etc.

In conclusion, the increase of I_1 and I_2 indicators could badly impact the process, making the global customer satisfaction decrease.

---

[8] We have implicitly supposed that a local performance increase determines a positive increase in the corresponding global indicator value.

Economic impact

The economic impact of an indicator strictly depends on the nature of the system investigated. The impact can be studied in relative terms, by comparing two different indicators operationalizing the same representation-target. In general, we cannot assert whether one indicator is economic or not, but we can only assert whether the indicator is more (or less) economic than another one.

To study and compare the economic impact of different indicators, we have to set up a mapping on the basis of their economic effects. Such a mapping cannot be defined only one way, but it depends on the nature of the process investigated. For instance, one of the most common mappings is based on the expenses of collecting information (see Fig. 4.10).

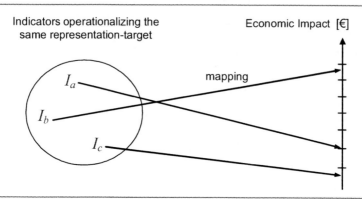

Fig. 4.10. Mapping performed to estimate the economic impact of a set of indicators

Example 4.18 A small company produces punched metal components. To check the quality of the manufactured holes, two possible indicators can be used:

I_1 *a diameter measurements*, taken by using an accurate calliper. To check each hole, the time needed is about 9 seconds.

I_2 *the result of a (go - no go) manual testing*, using a calibrated shaft of the minimum tolerable diameter. Time needed is about 3 seconds.

Supposing that the cost for measurements is directly proportional to time spent, then indicator I_2 can be considered three times more economical than indicator I_1.

Simplicity of use

This property, as the previous one, can be studied in relative terms, by comparing two (or more) different indicators operationalizing the same representation-target.

The comparison concerns the aspects related to *simplicity of use* (for example, indicators should be easy to understand, easy to use, they should have a clear meaning, they should be largely accepted, etc...).

Example 4.19 Likewise Example 4.18, we set-up a mapping to evaluate the simplicity of use of two indicators (I_1 and I_2), considering the following criteria:

(a) technical difficulty in performing measurements;
(b) time required.

These criteria are mapped on two corresponding 3-level ordinal scales (a_1: *low difficulty*, a_2: *medium difficulty*, a_3: *high difficulty*; b_1: *short time*, b_2: *medium time*, b_3: *long time*).

Global information on the *simplicity of use* can be determined, for example, through the sum of the level values. The smaller the sum, the simpler the indicator. Fig. 4.11 shows that the first indicator (I_1) requires more time and higher technical skills than the second one (I_2).

	(a) technical skill	(b) time required	simplicity of use (sum of the two level no.)
I_1	2	3	$(2 + 3) = 5$
I_2	1	1	$(1 + 1) = 2$

Fig. 4.11. Scheme to determine the simplicity of use of indicators I_1 and I_2. Indicators are defined in Example 4.18

It is interesting to note that scales (*a*) and (*b*) support the order property only, while the *simplicity of use* indicator also supports the interval property. This "promotion" has been introduced by the aggregation of indicators I_1 and I_2, by means of the sum of their respective level. As shown before, this sort of aggregation may sometimes produce paradoxical results

4.6.2 Properties of sets of indicators

A set of indicators is a way to represent a process, or a portion of it. Selected indicators should represent the real dimensions of a process, without omissions or redundancies. *Exhaustiveness* and *non-redundancy* (which are discussed in the following sections) are necessary but not sufficient conditions for this purpose.

Indicators set

A *set* or *family* of indicators is composed by the indicators selected to represent a generic process. These indicators can be grouped into *sub-sets*, depending on their characteristics.

State of a process

A generic process may lie in different conditions/states. The *state of a process* is the set of symbolic manifestations assumed by the indicators representing a specific process condition.

Example 4.20 Three indicators represent a company's sales:

I_1 *number of products daily sold*;
I_2 *daily turnover*;
I_3 *daily takings (not including the credit given)*.

Two possible process states are:

i-th day: $I_1(i) = 203$ pcs; $I_2(i) = 4820$ €; $I_3(i) = 3600$ €
j-th day: $I_1(j) = 178$ pcs; $I_2(j) = 5680$ €; $I_3(j) = 3546$ €

Each state is a "snapshot" of the process condition in a particular day.

Generally, complex processes can be structured according to different representation *dimensions*, for each of these it is possible to define (at least) one indicator. The selection of process *dimensions* is a difficult task, which has to be carefully performed by process *modellers*.

As stated before, empirical manifestations, distinguished according to the representation-target, should be related to distinguished symbolic manifestations (see Fig. 4.12).

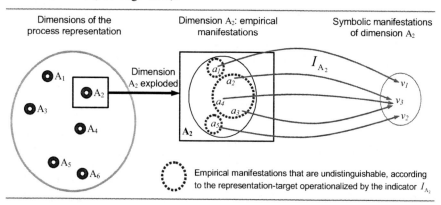

Fig. 4.12. Schematic representation of the concept of *indicators set*. For each process dimension (A_1, A_2, A_3, ...) it is possible to define one or more indicators. All the indicators form an indicators *set* or *family*. Indicator I_{A_2} represents the dimension A_2

Fig. 4.13 provides a schematic representation of two process states (state 1 and state 2).

Dimensions of the process representation-target

Empirical manifestations of the dimensions (in 2 particular states)

Corresponding symbolic manifestations (in 2 particular states)

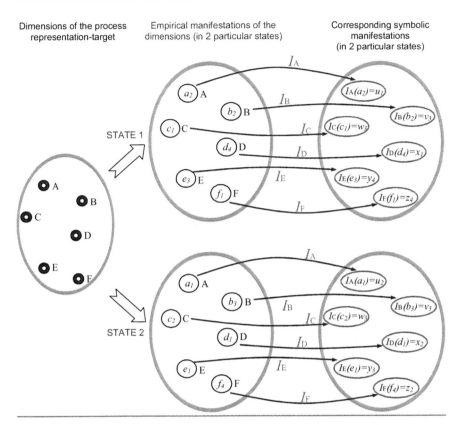

Fig. 4.13. Schematic representation of the concept of *state* of a process. Each *dimension* maps the empirical manifestations into corresponding symbolic manifestations.

Exhaustiveness

In establishing a *set* of indicators to represent a process, *exhaustiveness* is probably the most important property.

For a generic process modelled using indicators, the set of indicators is considered non-exhaustive when distinguishable empirical manifestations are mapped into the same symbolic manifestations.

The indicators *set* may be exhaustive only if indicators are well-defined (see Fig. 4.14).

The set of indicators is considered to be non-exhaustive in the following situations:

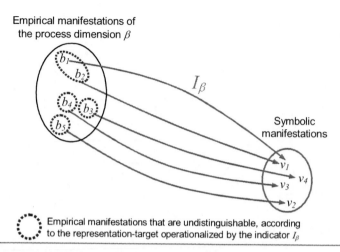

Empirical manifestations of
the process dimension β

I_β

Symbolic
manifestations

Empirical manifestations that are undistinguishable, according
to the representation-target operationalized by the indicator I_β

Fig. 4.14. Indicator correctly defined: distinguished empirical manifestations are mapped into distinguished symbolic manifestations

- One or more indicators are wrongly defined, because they do not map distinguishable empirical manifestations into separate symbolic manifestations (see Fig. 4.15).

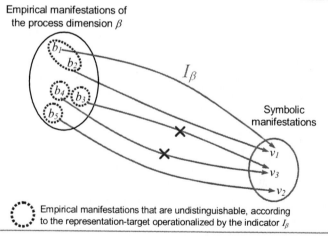

Empirical manifestations of
the process dimension β

I_β

Symbolic
manifestations

Empirical manifestations that are undistinguishable, according
to the representation-target operationalized by the indicator I_β

Fig. 4.15. Indicator wrongly defined: distinguished empirical manifestations are not distinguished by the indicator

- With reference to a representation-target, the model does not consider one or more process dimensions (see Fig. 4.16). In other words, the set is missing some indicators.

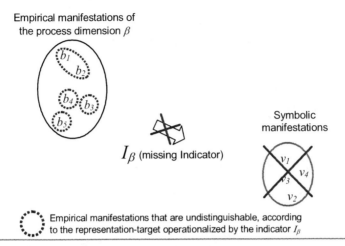

Empirical manifestations of
the process dimension β

I_β (missing Indicator)

Symbolic
manifestations

```
•'''•   Empirical manifestations that are undistinguishable, according
•...•   to the representation-target operationalized by the indicator $I_\beta$
```

Fig. 4.16. Missing indicators: with reference to a particular representation-target, the model does not consider one or more process dimensions

The property of exhaustiveness can be explained in another way. If indicators are unable to discriminate two process states – "*a*" and "*b*" – and if some empirical manifestations of the state "*a*" can be distinguished from these of the state "*b*", then the model is incomplete or inaccurate, and does not fulfil the property of exhaustiveness.

The condition may be tested by means of the following *formal test*:

<u>If</u> $\forall j \in F,\ I_j(a) \approx I_j(b)$

<u>and if</u>   empirical manifestations of the state "*a*" are distinguished by
            empirical manifestations of the state "*b*"

<u>then</u>   the indicators model is not exhaustive

being:
a and *b*   states of the process;
F   set (family) of indicators.

Example 4.21 A manufacturing company producing metal components, uses the following indicators:

I_1 *total number of units yearly produced*;
I_2 *manufacturing time*;
I_3 "*lead times*" (i.e. supply time, tool change time, etc...).

This set of indicators has been defined with the aim of differentiating the possible system conditions. If two possible states, undistinguished by the previous indicators, are distinguished by a further indicator – which has previously been ignored (for example I_4, the *number of defective units produced*) – then the set is not exhaustive:

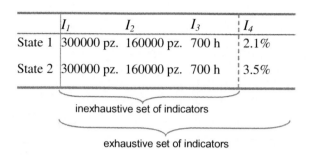

	I_1	I_2	I_3	I_4
State 1	300000 pz.	160000 pz.	700 h	2.1%
State 2	300000 pz.	160000 pz.	700 h	3.5%

inexhaustive set of indicators

exhaustive set of indicators

Fig. 4.17. In-exhaustive set of indicators (I_1, I_2, I_3), made exhaustive adding a new indicator (I_4)

Considerations on the exhaustiveness

Exhaustiveness is certainly the most important condition to guarantee consistency between indicators and the process represented (Roy and Bouyssou 1993). Exhaustiveness is fulfilled only when there are no process *states* contradicting it. So, the property testing criterion is linked to the principle of "inference".

Property verification needs a *modeller*, that is to say a subject with a deep knowledge of the examined process. The modeller should determine (1) if different process states can be distinguished in terms of empirical manifestations, and (2) if they are mapped into distinguished symbolic manifestations by the indicators in use. At present time, an automatic tool for this testing is not available.

Since every process is a dynamic system evolving over time, representation targets may change as time goes by. For that reason, every indicator, in order to be aligned with representation targets, need to be constantly modified or improved. Exhaustiveness is a practical tool to periodically check the consistency between representation targets and indicators (Flapper et al. 1996). If representation targets change, one or more indicators may not properly represent them, not satisfying the property of exhaustiveness.

The link between representation targets and firm strategy is provided by "accessory properties".

Non redundancy

If a set (or family) of indicators (F) is exhaustive, and if it continues to be exhaustive even when removing one indicator (I_k), then the removed indicator is redundant.

In formal terms:

<u>If</u>	F fulfils the property of exhaustiveness
<u>and if</u>	$\exists\ I_k \in F$: $F\backslash\{I_k\}$ still fulfils the property of exhaustiveness
<u>then</u>	I_k is a redundant indicator

where:

F	original set of indicators;
$F\backslash\{I_k\}$	original set of indicators, not including the indicator I_k.

Example 4.22 in a manufacturing company producing plastic component, the process is represented by four indicators:

I_1 total number of units (yearly) produced;
I_2 number of defective units (yearly) produced;
I_3 manufacturing time;

I_4 "efficiency of the production", calculated as: $I_4 = \dfrac{I_3 - I_5}{I_3}$

(term I_5 refers to "lead times", such as supply time, tool change time, repairing time, etc..);
I_5 "lead times".

Assuming that the set of indicators fulfils the property of exhaustiveness, the indicator I_3 is removed from the set. If the residual set (I_1, I_2, I_4, I_5) continues to be exhaustive, then the indicator I_3 is categorized as redundant (see Fig. 4.18).

if	I_1	I_2	I_3	I_4	I_5	is an exhaustive set of indicators
and if	I_1	I_2	$\cancel{I_3}$	I_4	I_5	is a set which continues to be exhaustive
then			$I_3 \longrightarrow$			is a redundant indicator

Fig. 4.18. Schematization of the concept of "redundant indicator"

Usually, indicators that can be deduced from other ones – that is, *derived* indicators, as in this case (I_3, which is a function of I_4 and I_5) – are redundant. The presence of redundant indicators does not provide additional information on the process.

In conclusion, with reference to the *Representation Theory*, an indicator is redundant when the empirical manifestations it maps are already considered by other indicators, or if it is scarcely significant for the representation of the process.

4.6.3 Properties of derived indicators

Derived indicators aggregate and summarize the information of a set of sub-indicators.

Generally, the more the process is complex, the more the indicators needed are numerous and different. Derived indicators simplify process analysing and monitoring.

Example 4.23 Likewise in Example 4.10, consider four *basic* indicators representing the concentrations of four different pollutants, to estimate the pollution level of the exhaust emissions of a motor vehicle.

The concentration of each pollutant is mapped into a 5 level scale by a single indicator. Let us suppose to analyse two possible conditions (*a* and *b*), in order to determine the worst (Fig. 4.19).

	I'_{NO_x}	I'_{SO_2}	I'_{CO}	$I'_{PM_{10}}$
Condition (a)	4	4	3	4
Condition (b)	1	1	1	5

Fig. 4.19. Comparison between two different pollution levels of the exhaust emissions of a motor vehicle

To evaluate global conditions, it is convenient to define a *derived* indicator (I), aggregating the information of the previous ones. I is defined as the maximum of the four "source" indicators values:

$$I(a) = \max\left\{I'_{NO_X}, I'_{HC}, I'_{CO}, I'_{PM_{10}}\right\} = \max\left\{4,4,3,4\right\} = 4 \tag{4.6}$$

$$I(b) = \max\left\{I'_{NO_X}, I'_{HC}, I'_{CO}, I'_{PM_{10}}\right\} = \max\left\{1,1,1,5\right\} = 5 \tag{4.7}$$

So, according to the derived indicator (I), the system condition (b) is worse than condition (a).

The aggregation of indicators can considerably simplify the analysis of the system, but it can also be questionable or misleading. The effectiveness of a derived indicator strongly depends on the aggregation rules (Franceschini et al. 2006). For instance, the condition (b) is considered the worst, even if the risk level of three pollutants (I'_{NO_2}, I'_{SO_2}, and I'_{CO}) is much lower than in condition (a). On the next sections we will illustrate some properties which may assist the aggregation of sub-indicators into derived indicators.

Property of monotony

Let us consider a set of sub-indicators aggregated by a *derived* indicator. If the increase/decrease of one sub-indicator is not associated to the increase/decrease of the derived indicator, then the derived indicator does not fulfil the condition of *monotony*.

This definition implicitly entails that the symbolic manifestations of the sub-indicators are represented using a scale with <u>order</u> relation. That is to say that it allows local comparisons among the symbolic manifestations, like $I_k(a) > I_k(b)$ (see Fig. 4.20). When indicators are represented on scales with no order relation (for example category scales: Yes-No, A-B-C, etc...), the property of monotony (as well as the concept of *local performance*) loses its meaning.

In more detailed terms, if a process is represented by different sub-indicators aggregated into a derived indicator (I_{TOT}), and if the process skips from state S to state S^*, increasing/decreasing one sub-indicator I_k, (not changing other indicators' performance), then I_{TOT} should increase/decrease too. Otherwise, I_{TOT} is not monotonous.

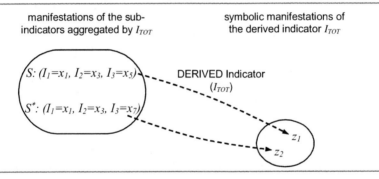

| manifestations of the sub-indicators aggregated by I_{TOT} | symbolic manifestations of the derived indicator I_{TOT} |

Fig. 4.20. Schematic representation of the condition of monotony. If process skips from state S to state S^*, being $I_1(S^*) = I_1(S)$, $I_2(S^*) = I_2(S)$, and $I_3(S^*) > I_3(S)$, then the Monotony entails that $I_{TOT}(S^*) > I_{TOT}(S)$

In formal terms:

<u>If</u>	$\forall j \in F \setminus \{I_k\},\ I_j(S^*) \approx I_j(S)$
<u>and if</u>	$I_k(S^*) > I_k(S)$
<u>and if</u>	$I_{TOT}(S^*) > I_{TOT}(S)$
<u>then</u>	the derived indicator I_{TOT} is monotonous

being:

F	indicators set (family);
I_k	increasing indicator;
$F \setminus \{I_k\}$	original set of indicators, not including I_k ;
I_{TOT}	*derived* (aggregated) indicator;
S and S^*	two process states.

Example 4.24 The pollution level of the exhaust emissions of a motor vehicle is estimated using the following model (see Example 4.10):

$$I_{TOT}^A = \max\left\{I'_{NO_X}, I'_{HC}, I'_{CO}, I'_{PM_{10}}\right\} \tag{4.8}$$

Assuming that pollution level skips from state S to state S^*, three sub-indicators do increase, the value of the derived indicator (I_{TOT}^A) not necessarily increases (see Table 4.3).

Table 4.3. Example of a non monotonous *derived* indicator (I_{TOT}^A)

	I'_{NO_X}		I'_{HC}		I'_{CO}		$I'_{PM_{10}}$		I_{TOT}^A
state S	1		1		1		3		3
state S^*	2	⇑ 3		⇑ 2		⇑ 3		3		3

In other terms, I_{TOT}^A can be "insensitive" to sub-indicators' variations.

The example shows that using a derived indicator which is not monotonous, we may lose some information (according to I_{TOT}^A, there is no difference between state S and state S^*). In Sect. 2.3 we have described some other indicators not fulfilling the property of monotony (*ATMO*, *AQI* and *IQA* indicators)

Property of compensation

The property of compensation can be studied when a process is represented by sub-indicators aggregated by a *derived* indicator. If changes of sub-indicators compensate each other – without making the *derived* indicator value change – then the derived indicator fulfils the property of compensation. In formal terms, a derived indicator (I_{TOT}) fulfils the property of compensation if the following condition is verified:

<u>If</u> $I_{TOT}(S^*) \approx I_{TOT}(S)$
<u>and if</u> $\exists\, I_i \in F: I_i(S^*) \neq I_i(S)$
<u>then</u> $\exists$ at least one indicator $I_j \in F: I_j(S^*) \neq I_j(S)$

being:
F original indicators set (family);
I_{TOT} *derived* (aggregated) indicator;
S and S^* two process states.

Example 4.25 With reference to Example 4.10, where the pollution level of motor vehicle exhaust emissions is estimated, we consider I_{TOT}^B as the synthesis indicator:

$$I_{TOT}^B = \frac{\left(I_{NO_X}^{"} + I_{HC}^{"} + I_{CO}^{"} + I_{PM_{10}}^{"}\right)}{4} \qquad (4.9)$$

As illustrated in Table 4.4, the pollution level skips from state S to state S^*. The decreases of I_{NO_X} and I_{HC} are compensated by the increase of I_{CO}. I_{TOT}^B value does not change.

Table 4.4. *Derived* indicator fulfilling the property of compensation

	I_{NO_X}		I_{HC}		I_{CO}		$I_{PM_{10}}$		I_{TOT}^B
state S	2		2		1		3		(2+2+1+3) / 4 = 2
state S*	1	⇓	1	⇓	3	⇑	3		(1+1+3+3) / 4 = 2

Sects. 2.2 and 2.3 show some examples of compensation of different derived indicators: *HDI* and *air quality indexes*. Compensation is a typical property of additive and multiplicative models.

4.7 Accessory properties

This book has illustrated many properties to support the analysis of indicators. However, before thinking of "how" to represent a particular aspect of the process, it is important to think of "which" process dimensions need to be represented. In practical terms, before defining process indicators, we should identify representation targets which are derived from the firm strategy. Indicators direct and regulate the activities in support of strategic objectives. Kaplan and Norton emphasize this link between strategies, action and indicators, considering four different perspectives (financial, customer, internal business process, learning and growth) (Kaplan and Norton 1996). Each perspective should be directly linked to reasonable representation targets. The following two accessory properties are introduced to help identifying representation targets which are consistent with the strategic objectives. We underline that the properties are defined "accessory" because they are helpful for testing process representation targets, rather than indicators.

- *Long term goals*. Since indicators should encourage the achievement of process' long-term goals, representation-targets should concern process' dimensions which are strictly linked to these goals.
- *Customer orientation*. In a competitive market, one of the main goals of every company is customer satisfaction. Many indicators focus on inter-

nal needs such as throughput, staff efficiency, cost reduction, and cycle time. While these needs are all laudable, they usually have little direct impact on costumers needs. So, it is important to identify process aspects with a strong impact on customer satisfaction. Quality Function Deployment is a valid tool to reach this objective (Franceschini 2001).

4.8   Indicators construction and check of properties

After illustrating major performance indicators properties, now we suggest an operative method for defining and testing the indicators of a generic process. The method – developed with more detail in Sect. 5 – is based on the following steps:

- definition of the *process* and identification of the characteristic *dimensions*;
- identification of representation-targets;
- analysis of the representation-targets time-horizon and impact onto process stakeholders (*accessory properties* testing);
- preliminary definition of indicators;
- for each indicator, check of the *consistency with the representation-target*;
- indicators selection and check of exhaustiveness and *non redundancy* properties;
- definition of the measuring scale and definition of the data collecting procedure for each indicator; general properties testing (simplicity of use, economic impact, level of detail, non counter-productivity, ...);
- check of derived indicators properties: *monotony* and *compensation*.

This methodology is based on a "top-down" testing. First, representation-target should be identified in order to be consistent with firm strategies (*accessory properties*). Then, a preliminary definition of process indicators is given. For each indicator, we should make sure it represents a particular process representation-target (*consistency with the representation-target* property). Next step is in testing the properties of the indicators set (*exhaustiveness, non redundancy*), then other properties of single indicators are tested (*general properties: level of detail, non counter-productivity, economic impact, simplicity of use*). Basically, before evaluating single indicators in detail, it is prior to assess that indicators are well integrated each other. In this phase, one of the major difficulties is identifying or predicting all possible process states.

After testing indicators "general properties" we should check derived indicators properties ("monotony" and "compensation"), and the rules with which sub-indicators are aggregated into derived indicators.

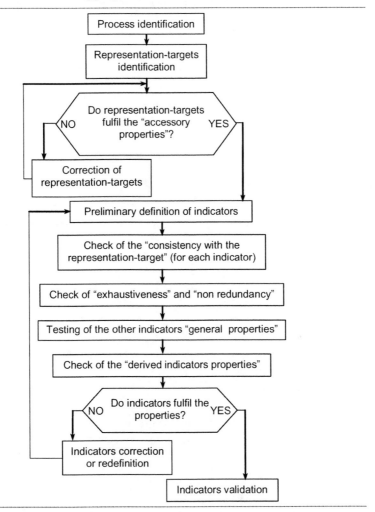

Fig. 4.21. Scheme of operational methodology

As illustrated in Fig. 4.21, the procedure requires several recursive steps (definition, test, correction, redefinition, and so on..) before developing a proper model. Analysing indicators from a formal mathematical perspective makes it easier identifying possible drawbacks.

The proposed approach mainly focuses on indicators testing rather than an indicators designing. This methodology contributes to making aware of the risk of defining/selecting improper indicators and – in our opinion – it may be also used for integrating other existing approaches.

Example 4.26 An *automotive* company establishes a system of indicators to evaluate the performances of a new car.

• *Process Identification.*
The most important process *dimensions* are:

1. technical features;
2. running dynamics performances;
3. fuel consumption;
4. polluting emissions;
5. comfort;
6. cost of materials (raw materials, semi-processed products, components purchased);
7. production costs.

• *Identification of representation-targets.*
For each process dimension we should identify the representation-targets (see Table 4.5).

Table 4.5. List of the major representation-targets related to the process dimensions

Process dimension	Representation-target
1. Technical features	1.1 - Car weight 1.2 - Engine characteristics 1.3 - Passenger compartment volume
2. Running dynamic performances	2.1 - Maximum engine power 2.2 - Starting acceleration 2.3 - Cornering (entering a bend) 2.4 - Cornering (coming around a bend) 2.5 - Braking performances
3. Consumption	3.1 - Average fuel consumption evaluated on three standard tracks (urban/suburban/mixed)
4. Polluting emissions	4.1 - Pollution level of the car exhaust emissions
5. Comfort	5.1 - Car seats comfort 5.2 - Vibrations 5.3 - Noise
6. Cost of materials	6.1 - Cost of base materials 6.2 - Purchase cost of finished components (for instance car seats, brakes, air-conditioner, etc.)
7. Production costs	7.1 - Manufacturing cost (for instance chassis, front-end, doors construction costs, varnishing cost, etc.) 7.2 - Car assembly cost 7.3 - Handling charges
...	...

- *Representation-targets analysis and testing.*

Representation-targets should be analysed and tested according to the "accessory properties" (see Table 4.6).

Representation-targets should be consistent with the accessory properties. The condition is fulfilled, after redefining/adding/removing some of them.

Table 4.6 Accessory properties for the representation-targets testing

a - representation-targets time horizon	a_1 - the car vehicle should have high performances and should not be subjected to quick deterioration;
	a_2 - the product should fulfil the current (and future) regulations about the polluting emissions;
	a_3 - in the short term, the product cost should be competitive, in order to penetrate the market.
b - impact onto process stakeholders	b_1 - car performances should be competitive;
	b_2 - the car design should focus on passengers safety and comfort.

- *Preliminary definition of indicators.*

The next step consists in defining indicators (Table 4.7). Selected indicators include a *derived* indicator (4.1.5) and a *subjective* indicator (5.1.1). Most of them are measurements, except for these ones, and those referred to cost estimations (6.1.1 ÷ 7.3.1).

It is essential to test the indicators *consistency with the representation-targets*. The question to be asked is: "*Are representation-targets properly operationalized by selected indicators?*".

- *Testing of indicators sets.*
  - After indicators selection, we should test the property of *exhaustiveness*. Each process *dimension* is separately analysed. Considering the first dimension – *Technical features* – and the first representation-target – *car weight* – the indicator 1.1.1 (*vehicle mass in conditions of medium load*) is not exhaustive (for example, if we want to consider also the vehicle mass in full load conditions). In this case, indicator 1.1.1 may be modified or completed by new indicators.

    Furthermore, still concerning the first dimension, the selected indicators fail to consider the vehicle external dimensions (length, height, width) or other important features such as the engine type (petroleum, supercharged, *common-rail* diesel, diesel with pump injector, etc..). As a consequence, new representation-targets and new indicators are added (see Table 4.8).

    This procedure extends to all the process dimensions, and representation-targets.
  - *Non redundancy* should be tested analysing together the indicators of the *set*. For example, if there is a correlation between the indicator 5.2 (*vibrations*) and the indicator 5.3 (*noise*), one of them may be classified as redundant and

then removed from the model. Generally, correlations are limited to indicators representing the same process dimension.

Table 4.7. Preliminary list of the selected indicators

Representation-target	Indicators
1.1 - Vehicle mass 1.2 - Engine characteristics 1.3 - Passenger compartment volume	1.1.1 - Vehicle mass in average load conditions 1.2.1 - Engine cubic capacity [cm^3] 1.3.1 - Internal compartment volume measurement [dm^3]
2.1 - Maximum engine power	2.1.1 - Maximum engine power [kW], measured through bench experimental test (in controlled conditions).
2.2 - Starting acceleration	2.2.1 - Average vehicle acceleration from the speed of 0 to 100 km/h.
2.3 - Cornering (entering a bend)	2.3.1 - Maximum speed of the vehicle entering a bend with definite radium, in safety conditions (defined by specific standards).
2.4 - Cornering (coming around a bend)	2.4.1 - Maximum speed of the vehicle coming around a bend with definite radium, in safety conditions (defined by specific standards).
2.5 - Braking performances	2.5.1 - Braking space for the complete vehicle stop from the speed of 100 km/h in normal conditions (dry road, tyres with medium wear, and medium car load; 2.5.2 - Average vehicle deceleration, in the same test condition.
3.1 - Average fuel consumption	3.1.1 - Average distance covered [km] to the litre, on 3 standard tracks (urban/suburban/mixed)
4.1 - Pollution level of the car exhaust emissions	4.1.1÷4.1.4 - Concentration of each pollutant. The indicators are similar to those of the Example 4.10. 4.1.5 - Derived indicator to "summarize" the global polluting level.
5.1 - Car seats comfort	5.1.1 - Indicator of customer satisfaction. The indicator is obtained from a customers' interview about car seats comfort.
5.2 - Vibrations	5.2.1 - Measurement of the vehicle internal vibrations (amplitude and frequency), in predefined running conditions.
5.3 - Noise	5.3.1 - Phonometric measurement of the vehicle internal maximum noise (in dB), in predefined running conditions. Measured values are compared to standard reference values, or to the competitors'.
6.1 - Cost of the base materials	6.1.1 - Estimation of the raw material costs (with reference to a single vehicle).
6.2 - Purchase cost of the finished components (for instance car seats, brakes, air-conditioner, etc.)	6.2.1 - Estimation of the finished components cost (with reference to a single vehicle).
7.1 - Manufacturing cost (for instance chassis, front-end, doors construction costs, varnishing cost, etc.)	7.1.1 - Estimation of the total manufacturing cost (with reference to a single vehicle).
7.2 - Car assembly cost	7.2.1 - Estimation of the assembly cost (with reference to a single vehicle).
7.3 - Handling charges	7.3.1 - Estimation of the average material handling cost (with reference to a single vehicle).
...	...

Table 4.8. Additional indicators to represent the first process dimension (*technical features*)

Representation-target	Indicators
1.2 - Engine type	1.2.2 - engine type: petroleum, supercharged, common-rail diesel, diesel with pump injector, etc.
1.4 - Vehicle external dimensions	1.4.1 - vehicle length; 1.4.2 - vehicle width; 1.4.3 - vehicle height.
...	...

- *Testing of the single indicator properties.*
  Continuing with the procedure, we must consider the other properties of single indicators (*simplicity of use, economic impact, non counter-productivity*, etc.). For example, regarding the indicators 1.4.1 (*vehicle length*), 1.4.2 (*vehicle width*), 1.4.3 (*vehicle height*), we can consider the accurateness of the measuring instruments.
  - *Economic impact* and *easiness of use* are usually evaluated on the basis of empirical comparisons. For example, considering the economic impact of the indicator 2.1.1 (*maximum engine power*), it is generally more practical to measure it through experimental bench tests (in controlled conditions), rather than through testing on the road.
  - Testing the indicators *non counter-productivity* is quite difficult. Each indicator should be analysed to find whether it can be counter-productive in terms of process general performance.
    For instance, the indicator 2.1.1 (*maximum engine power*) should not increase too much, or it will worsen other aspects, for example *cornering* performance (2.3 and 2.4). In fact, when the engine power is excessive, it can not be completely conveyed to the ground and this could have dangerous consequences (understeer/oversteer). Additionally, indicator 2.1.1 could be counter-productive for other process dimensions, such as the internal *noise* and *vibrations* (5.2 and 5.3).

- *Testing of derived indicators.*
  For each *derived* indicator (for example the indicator 4.1.5 indicating the global polluting level) it is possible to test the properties of *monotony* and *compensation*.

The proposed operational method support the selection and the testing of the indicators during the design activities. In the next chapter it will be dealt with the problem of the building and designing a performance measurement system.

5. Designing a performance measurement system

5.1 Introduction

The present chapter discusses how to establish and maintain a performance measurement system. This operation is at the "heart" of performance-based management processes. A performance measurement system, flowing from the organizational mission and the strategic planning process, provides the data that will be collected, analyzed, reported, and, ultimately, used to make sound business decisions.

This chapter is divided into nine sections, including the present introduction.

Sects. 5.2 and 5.3 present the basic concepts and components of an integrated performance measurement system. The fourth section discusses some of the major approaches to develop performance measurement systems. Incidentally, we will analyse the following methods: the *Balanced Scorecard*, the *Critical Few*, the *Performance Dashboards*, and the EFQM award (*European Foundation for Quality Management*). Description is supported by the use of practical examples. The fifth, sixth and seventh sections discuss how to synthesise and develop indicators, and how to maintain an effective performance measurement system. In conclusion, the two last sections are about possible misuse of indicators and their impact onto organized systems.

5.2 The concept of performance measurement system

The concept of performance measurement is straightforward: you get what you measure, and you cannot manage a system unless you measure it.

In the document *Performance Measurement and Evaluation: Definitions and Relationships* (GAO/GGD-98-26), the U.S. General Accounting Office (GAO) provides the following definition: "*Performance measurement is the ongoing monitoring and reporting of program accomplishments, particularly progress towards pre-established goals. It is typically conducted by program or agency management. Performance measures may address the type or level of program activities conducted (process), the direct products and services delivered by a program (outputs), and/or the results of those products and services (outcomes). A "program" may be any activity, project, function, or policy that has an identifiable purpose or set of objectives".*

Performance measures are tools to understand, manage, and improve organization activities. The effective performance measures allow us to understand:

- how well we are doing (correct process representation);
- if we are meeting our goals (identification of the goals and the reference standards);
- if our customers are satisfied (control of the process development);
- if our processes are in control (control organization effectiveness and efficiency parameters);
- if and where process improvements are necessary (identification and correction of problems).

A performance measure (or indicator) is composed of a number and a unit of measure. The number gives us a magnitude (how much) and the unit gives the number a meaning (what). Performance measures are always tied to a representation-target. In Chap. 3 we illustrated the reason why measurements can be considered a subset of indicators.

As already seen in Sect. 1.4, most performance indicators may be related to the following three types:

- *effectiveness*: a process characteristic indicating the degree to which the process output conforms to requirements (*Are we doing the right things?*);
- *efficiency*: a process characteristic indicating the degree to which the process produces the required output at minimum resource cost (*Are we doing things right?*);
- *customer care*: the degree to which the process users/customers appreciate the provided performances.

5.2.1 Why performance measurements?

Here we briefly introduce some reasons why to adopt a performance measurement system:

- performance measurement provides a structured approach for focusing on a program's strategic plan, goals, and performance;
- measurements focus attention on what is to be accomplished and compels organizations to concentrate time, resources, and energy on achievement of objectives. Measurements provide feedback on progress toward objectives;
- performance measurement improves communications internally among employees, as well as externally between the organization and its customers and stakeholders. The emphasis on measuring and improving performance (*results-oriented management*) creates a new climate, affecting all the organizations aspects;
- performance measurement helps justify programs and their costs. Measurements provide the demonstration of a program's good performance and sustainable impacts with positive results, in order to support the decision making process.

5.2.2 What performance measures won't tell you

Even though the performance measurement system is a valuable tool in managing and controlling process development, it is not able to tell you about the following:

- *The cause and effect of outcomes are not easily established.* Outcomes can, and often do, reveal the impact of the program, but without collaborating data, it is difficult to demonstrate that your program was the cause of the outcome(s). The outcomes of a methodology are inevitably affected by many events outside control. Another conditioning element is the time difference between cause and effect.
- *Poor results do not necessarily point to poor execution.* If the performance objectives are not being met, it is obvious that something is wrong, but performance information does not always provide the reason. Instead, it raises a flag requiring investigation. Possibilities include performance expectations that were unrealistic or changed work priorities.
- *Measurements are only a model to the actual system reading.* The measured system is not the same as the actual system. It is only an approximate of it. So, the level of detail mainly depends on the model.
- *Performance measures do not ensure compliance with laws and regulations.* Performance measures do not provide information on adherence to laws and regulations or the effectiveness of internal controls. For example, a building could be constructed more quickly if safety controls

and funding limitations were ignored. Because compliance and internal controls often have a direct effect on performance, care should be taken to supplement performance measurement. This could be done with other oversight activities, to ensure that controls are in place and working as intended and that activities are adhering to laws and regulations.

5.2.3 Major difficulties in implementing a measurement systems

Brown's quote well synthesizes the possible problems in the construction of a performance measurement system: *"The most common mistake organizations make is measuring too many variables. The next most common mistake is measuring too few"* (Brown 1996). In general, the most common difficulties are:

- amassing too much (or too little) data. Consequently, data may be ignored or used ineffectively;
- focusing on the short-term. Most organizations only collect financial and operational data, forgetting to focus on the longer-term measures;
- collecting inconsistent, conflicting, and unnecessary data (Flapper et al. 1996; Schmenner et al. 1994). All data should lead to some ultimate measure of success for the company. An example of conflicting measures would be measuring reduction of office space per staff, while, at the same time, measuring staff satisfaction regarding the facilities.
- measures may not be linked to the organization's strategic targets;
- inadequate balancing of the organization's performances. For instance, the manager of a restaurant may have perfect kitchen efficiency (the ratio of how many dishes sold to the amount thrown away) by waiting until the food is ordered before cooking it. However, the end result of his actions dissatisfied customers, because of the long wait (see the concept of *exhaustiveness*, in Sect. 4.6.2);
- measuring progress too often or not often enough. There has to be a balance here. Measuring progress too often could result in unnecessary effort and excessive costs, resulting in little or no added value. On the other hand, not measuring progress often enough puts you in the situation where you don't know about potential problems until it's too late to take appropriate action.

5.3   The construction process

Generally speaking, processes can be considered like natural organisms evolving over time and influenced by the environment in which they live. The process manager defines the process targets and the corresponding performance indicators. All interested parties should be actively involved in the process, knowing the process targets, how they contribute to the process success, and the stakeholders' measurable expectations (Neely et al. 1997).

To establish a performance measurement system three basic aspects should be considered:

- the strategic plan;
- analysis of the key sub-processes;
- stakeholder needs.

5.3.1 The strategic plan

Strategic plan sets the foundation for effective performance measurement systems. Traditional performance measurement systems that focus on the wrong set of performance measures can actually undermine an organization's strategic mission by perpetuating short-sighted business practices. For this reason, it is appropriate to discuss the critical elements of strategic plans and review the compatibility of strategic plans; to an integrated performance measurement system. A well-developed strategic plan should contain the basic information necessary to begin the formulation of an integrated performance measurement system as shown in Table 5.1.

With performance measurements collected from the strategic plan, we should determine the quality of information and current use of existing indicators. The objective is to find out which indicators are maintained and monitored, and who are the owner(s) and data customer(s). Answering the following five questions should provide enough information for this step:

- What information is being reported?
- Who is responsible for collecting and reporting performance information?
- When and how often is the performance measure reported?
- How is the information reported?
- To whom is the performance measure reported to?

Mapping performance measures to the strategic plan can be performed by using a spreadsheet or a table, as those used for the *Quality Function Deployment* methodology (Franceschini 2002).

Table 5.1. Strategic plan element and performance measurement attributes

Strategic plan elements	Performance measurement attributes
Strategic Goal	Articulates the enduring mission or "end state" desired.
Objective	Describes the strategic activities that are required to accomplish the goal.
Strategy	Defines strategic (long-term) requirements that link to objectives. Typically contain dates, basis of measurement, and performance aspirations (targets).
Tactical Plans	Identifies the short term requirements that link to strategy. Typically contain cost, time, milestone, quality, or safety attributes as well as performance targets.

Fig. 5.1 shows an example of mapping. Targets are organized according to hierarchical criteria (targets tree), and summarized in the left of the so-called *Relationship Matrix*. Performance measures (indicators) are placed on the top of the Relationship Matrix (columns).

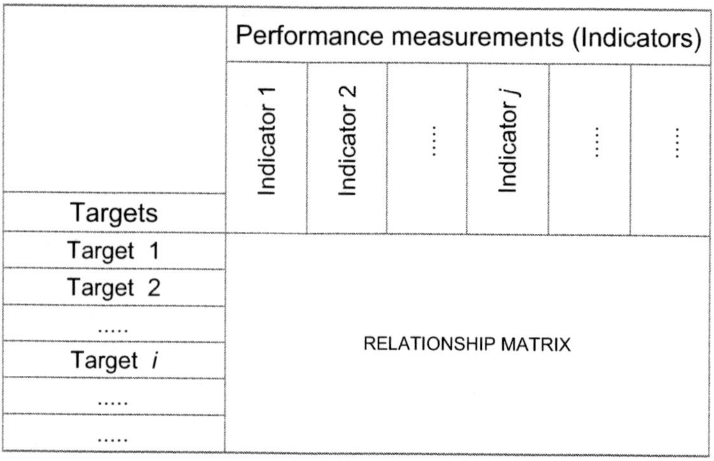

Fig. 5.1. Example of performance measurements mapping. Targets linked to the strategies are listed on the left of the Relationship Matrix (rows). Performance measures (indicators) are shown on the top of the Relationship Matrix (columns)

Usually targets may "impact" different performance measurements, and *vice-versa*. This methodology offers a way of unravelling this complex network of relationships through the use of the Relationship Matrix, $\mathbf{R} \in \Re^{m,n}$ (being m the targets number and n the indicators number).

Relations between "targets" and "performance measurements" are represented by specific symbols to indicate "weak", "medium", or "strong" relationships, respectively. Commonly used symbols are: a triangle for weak relationships, a circle for medium relationships, and two concentric circles for strong relationships (Fig. 5.2).

If no relationship is apparent, the corresponding intersections in the matrix are left blank. Rows or columns left completely blank indicate zones where the transformation of "targets" into "performance measurements" is inapplicable. This methodology makes it possible to transform targets into control actions, due to the very fact that it includes repeated cross-checks on the various analysed aspects.

By way of example, Fig. 5.2 shows the extract of a Relationship Matrix related to a Help Desk service for the ICT (*Information Communication Technology*) function of an important broadcasting company (DISPEA 2005).

		Performance measurements								
Δ: weak relationship O: medium relationship ●: strong relationship		Routing effectiveness	Reply accuracy	Uniformity of replies (from different operators)	Time for requests implementation	Competence perception	Answered calls percentage	Courtesy of responses	Security of data keeping	Number of active lines
Targets	Importance									
Reliability	5	●	O	●	O	Δ			O	
Responsiveness	5					●	Δ			
Competence	4	●	O	●		●				
Access	4						●			O
Courtesy	3			Δ				●		
Communication	3		Δ	●		Δ		O		
Credibility	3	Δ		O		O				
Security	5	Δ							●	
Understanding/ Knowing the customer	4	O	Δ	Δ		Δ		O		
Tangibles	3									●
Present model values		93%	A	A	22 min	MB	98%	M	3%	3
Target values		>90%	A	A	20 min	MB	>99%	A	<5%	5

Fig. 5.2. The Relationship Matrix for a Help Desk service. Relations between "targets" and "performance measurement" are coded by specific symbols (DISPEA 2005)

Parallel to the "measurements" axis, on the bottom line of the matrix, a third area is brought into focus, the axis of the "performances target values". They constitute specific reference values, that serve as guidelines for the planning of the performance measurement system.

Considering the example in Fig. 5.2, service *representation-targets* are selected using the model suggested by Parasuraman, Zeithaml and Berry (PZB model) – one of the most accredited model in literature for quality service evaluations. The model identifies 10 key elements for the service quality, called "determinants" (see Table 5.2) (Parasuraman et al. 1985, 1988, 1991; Franceschini 2001).

Table 5.2. Determinants for the service quality, according to the PZB model (Parasuraman et al. 1985). With permission

Determinant	Description
Reliability	It involves consistency of performance and dependability; it means that the firm performs the service right the first time; it also means that the firm honours its promises.
Responsiveness	It concerns the willingness or readiness of employees to provide service; it involves timeliness of service.
Competence	It means possession of the required skills and knowledge to perform the service.
Access	It involves approachability and use of contact.
Courtesy	It involves politeness, respect, consideration, and friendliness of contact personnel.
Communication	It means keeping customers informed in language they can understand and listening to them.
Credibility	It involves trustworthiness, credibility, and honesty; it involves having the customer's best interest at heart.
Security	It is the freedom from danger, risk, or doubt.
Understanding/ Knowing the Customer	It involves making the effort to understand the customer's needs.
Tangibles	They include the physical evidence of the service.

Each determinant (first level item) may be "exploded" in more detailed sub-items, depending on the examined process (see Table 5.3).

Process targets differently affect the manager's degree of satisfaction; so they must be ranked. The classical method solves this problem, associating each target with a numeric (importance) value, for example 1 (for a requisite of negligible importance) to 5 (for an indispensable requisite) (see Fig. 5.2). The most common and effective techniques to identify the service targets are based on personal interviews, or the so-called focus groups (Franceschini 2002).

Table 5.3. "Explosion" of three determinants ("reliability", "responsiveness", "competence") into a list of second level items (DISPEA 2005)

Determinant	Second level item
Reliability	Service pricing accuracy. Service punctuality. Promptness in phoning the customer back.
Responsiveness	Promptness in fixing appointments. Promptness in forwarding the documentation to the customer. Skill and knowledge of the front-office operators.
Competence	Skill and knowledge of the back-office operators. Service pricing accuracy.

Table 5.4. List of indicators selected for a Help Desk service monitoring (see Fig. 5.2) (DISPEA 2005)

Indicator	Definition	Measuring scale
Routing effectiveness	Percentage of calls switched less than 3 times.	%
Reply accuracy	Response relevance in terms of problem-solving, exhaustiveness, completeness, and consistency with correct procedures.	HIGH / MEDIUM / LOW
Uniformity of replies	Degree of uniformity of different operators.	HIGH / MEDIUM / LOW
Time for requests implementation	Average time before putting a customer request into action.	Minute
Competence perception	Customer's perception towards the operators' competence in providing responses.	POOR / SUFFICIENT / GOOD / VERY GOOD / EXCELLENT
Answered calls percentage	Ratio between the number of answered calls and the total amount of calls directed to the help desk.	%
Courtesy of responses	Operator's courtesy in answering the customers.	HIGH / MEDIUM / LOW
Security of data keeping	Ratio between the number of customers' complaints for some data lost, and the total number of calls.	%
Number of active lines	Number of available telephone lines that can be used at the same time.	Number

Following this, "targets", may be translated into "performance measurements". Indicators identified must "cover" all the targets, not neglecting any relevant aspect. For each indicator, it is necessary to define measuring scale, data collecting frequency and procedure, type of representation (through tables, histograms, historical series etc...). Table 5.4 illustrates the list of indicators related to the previous example (Table 5.2).

5.3.2 Identification of the key sub-processes

Processes are the implementation of the strategic targets. Each structured organization is composed by several sub-processes, which differently impact targets. To be effective a performance measurement system should identify sub-processes, which mostly "influence" targets. All process activities are divided into sub-processes and organized into a hierarchy, depending on their impact onto targets. This activity is performed by the use of the so-called "process maps", containing the qualitative and quantitative important information. Process-maps are graphic representations of activities, the interfaces, the flow of information and the responsibilities connected to the various process actors. These maps provide a detailed "snapshot" of an organization.

The methodology is structured into the following stages:

1. preliminary analysis of the processes;
2. drafting of the process maps;
3. analysis of the process maps.

Preliminary analysis of the processes

The purpose of this stage is to obtain a general knowledge of the process framework, focussing attention onto the aspects relating to customers (in terms of facilities and personnel).

The fundamental steps are:

- identification of the organization activities;
- identification of the subjects which the organization interfaces with;
- identification of the subjects which customers/users interface with;
- identification of the information to be managed, time required, possible problems or critical aspects.

This stage is essential, since it identifies the information necessary to delineate the organizational context. Furthermore, it can also be useful to reorganize and restructure the current process organization.

Drafting of the process maps

A generic process can be broken down into further sub-process. The breakdown may be performed at different levels, as illustrated in Fig. 5.3.

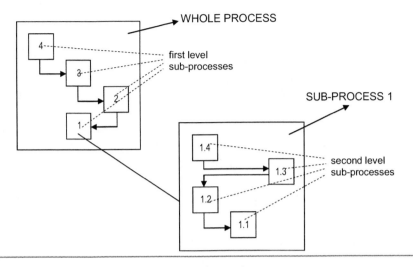

Fig. 5.3. Example of process breakdown into sub-processes

The target level of detail depends on different aspects:

- the "parent" process' complexity;
- deeper levels should correspond to basic procedures – represented by precise performance indicators – in which the actors' responsibilities are clearly defined;
- the number of levels should not be higher than 4 or 5, not to complicate the analysis.

Fig. 5.4 shows an example of the process map.

Phase	Procedure	Volume	Time	Database access	User	Start/ End	Responsibilities		
							Actor 1	Actor 2	Actor ...
1	description of procedure 1			data base	🧍	input →	procedure 1		
2	description of procedure 2			data base					procedure 2
3	description of procedure 3				🧍			procedure 3	
...	description of procedure ...			data base	🧍	output ←	procedure ...		

Fig. 5.4. General scheme of a process map

The first action in the construction of a process map consists in defining the following elements:

- process *input*;
- procedures/activities (represented by blocks in the process map);
- process *output* (results).

Afterwards, the actors' responsibilities are defined, identifying "who does what". So it is possible to analyse the information flow, through different organization activities.

The process map includes information about the interface between organization and customers, the amount of data exchanged, the (possible) presence of informative systems, the time required for activities development.

For a more detailed description of the process, the process map is completed with the following documents:

- references to the organization existing practices;
- a short description of the process salient phases;
- information on the process average time (the sum of times for single activities must equal the process total time);
- detailed description of the process critical aspects, and the possible causes.

Analysis of the process maps

The last stage of the methodology consists in analysing process maps. The purpose is to determine process' efficiency and effectiveness, defining *how*, *where*, and *when* performing the Quality monitoring.

Furthermore, it is necessary to identify and to patch up the process problems and critical aspects slowing the activities down.

Process activities performed by different organization functions can be separated by means of a process "vertical" reading. A virtual superimposition of the process maps may be useful to identify the organization functions workload and involvement.

Finally, we should identify process "vicious circles", trying to rationalize process activities and procedures.

5.3.3 Stakeholder needs

Stakeholders is a common term in performance-based management referring to those people who have a stake in the future success of an organization. It is imperative to have a very clear idea of who these people are, and *what* are their needs and expectations. Usually, if they have a share in the

process results, they play an active role in the activities development (Atkinson 1997).

A strategy to systematically understand what these stakeholders want
and expect has to be developed. However, depending on the stakeholder,
different techniques or tools are used. For customers, organizations often
use surveys or customer focus groups. For employees, surveys, focus
groups or discussions are excellent tools.

Developing performance measurement information based on stakeholders serves two purposes:

- it evaluates whether tactical plans, such as customer satisfaction and
  employee commitment, are being met;
- it provides a means of testing the presumed cause and effect relationships between performance measures and strategies. For example, does
  higher quality result in increased sales?

In many organizations, leadership commitment to the development and
use of performance measures is a critical element in the success of the
performance measurement system. Four specific ways to positively impact
the involvement of Management are (Thomson and Varley 1997):

- *Delegate responsibilities and empower employees.* Involvement creates
  ownership. It increases employees' loyalty, commitment and accountability.
- *Develop good communication processes.* A good communication process provides a critical link between the tasks employees perform and
  the corporate strategic plan/measures.
- *Always seek feedback.* Managers need to know what employees think
  about their jobs and the company - especially if they are not in a-
  lignment with the company's strategic direction. Doing so creates accountability for both employees and senior management.
- *Responsibilities definition.* Each performance measure needs to have an
  owner who is responsible for that measure. Employees need to know
  how the measurement(s) for which they are being held accountable relates to the overall success/failure of the organization.

5.3.4 Vertical and horizontal integration of performance measures

Performance measures need to be integrated in two directions: vertically
and horizontally. Vertical integration of performance measures motivates
and improves operating performance by focusing all employees' efforts on

the organization's strategic objectives. On the other hand, horizontal alignment assures the optimization of work flow across all process and organizational boundaries. Fig. 5.5 provides a simplistic sample of how different levels of indicators are deployed at different levels within the organization.

Characteristics of vertically integrated performance measurements include:

- the indicators' time alignment;
- stating the target values to accomplish;
- integration between process measurements and results measurements;
- definition of the responsibilities at each level of the organization framework.

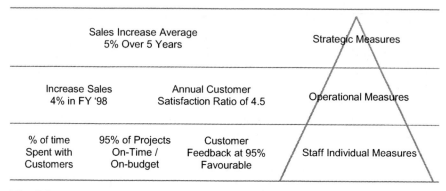

Fig. 5.5. An example of how different levels of indicators are deployed at different levels within the organization (Performance-Based Management Special Interest Group, Vol. 2 2001). With permission

From the customer's viewpoint, a company may be seen as a perfectly structured body, with no boundaries between activities and/or functions. Customers do not "see" process boundaries through which their products flow, but they care about the attributes of the product delivered to them.

An excellent example of this concept of horizontal alignment of performance measurements is the procurement cycle. The procurement organization may measure cycle times to improve customer satisfaction with the procurement process. However, the procurement requester may see the cycle time for procurements as much more than the procurement organization's portion of that cycle. From the procurement requester's viewpoint, all of the processes (beginning with the identification of the need for a product or service to the actual delivery of the product or service) represent the complete procurement cycle. To capture the entire cycle, many departments (computer services, mail room, procurement, receiving, property management and transportation) might all need to be involved.

5.4   A review of the major reference models

When you are developing a performance measurement system, you should consider a conceptual reference model. Experience has shown that a reference model is particularly important when you are beginning to develop a measurement system for the first time. In literature, you can find different approaches. In the following sections we provide a short description of some of them.

5.4.1 The concept of "balancing"

The concept of *balancing* performance measurements took root in 1992 when Robert Kaplan and David Norton – from Harvard University – first introduced the *Balanced Scorecard* methodology. The gist of this concept is to translate business mission accomplishment into a *critical set* of measures, distributed among an equally critical and focused set of business perspectives (*dimensions*). Through time, many variations of the concept will have surfaced. This will be due mainly to the fact that no two organizations are alike and their need for balanced measures and business perspectives vary. Regardless, the two key components of all of these frameworks are a balanced set of measures and a set of strategically focused business perspectives.

This concept is the starting point of several operational approaches:

- the Balanced Scorecard method;
- the Critical Few method;
- the Performance Dashboards;
- the EFQM (European Foundation for Quality Management) model, directly linked to the Malcolm Baldridge National Quality Award (American) model.

They will be presented and discussed in the following sections.

5.4.2 The "Balanced Scorecard" method

In 1992, Robert Kaplan and David Norton introduced the *Balanced Score-card* concept as a way of motivating and measuring an organization's performance (Kaplan and Norton 1992).

The concept takes a systematic approach in assessing internal results, while probing the external environment. It focuses as much on the process of arriving at "successful" results, as on the results themselves.

Indicators that make one *dimension* look good while deflating another are avoided, thus minimizing negative competition between individuals and functions. This framework is intended for top managers in an organization to be able to obtain a quick and comprehensive assessment of the organization in a single report. Use of the *Balanced Scorecard* requires executives to limit the number of measures to a vital few, allowing them to track whether improvement in one area is being achieved at the expense of another area.

The method looks at four interconnected perspectives (*dimensions*). These are:

- *Financial* – How do we look to our stakeholders?
- *Customer* – How well do we satisfy our internal and external customer's needs?
- *Internal Business Process* – How well do we perform at key internal business (sub)processes?
- *Learning and Growth* – Are we able to sustain innovation, change, and continuous improvement?

A graphic representation of these perspectives is provided in Fig. 5.6.

The *Balanced Scorecard* provides a way for management to look at the well-being of their organization from the four identified perspectives. Each perspective is directly linked to performance targets. In this framework, customer satisfaction drives financial success; effective and efficient business processes ensure high levels of customer satisfaction; and sustained, continuous improvement enhances the organization's operational performance. Each perspective also has a secondary influence, as shown in Fig. 5.6.

A way to identify performance indicators for each of these perspectives is to answer the following questions:

- *Financial* – What are our strategic financial objectives?
- *Customer* – What do we have to do for our customers in order to ensure our financial success?

- *Internal Business Process* – Which of our business processes most impact customer satisfaction?
- *Learning and Growth* – What improvements can be made to ensure sound business processes and satisfied customers?

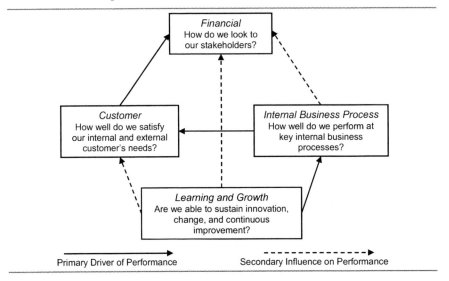

Fig. 5.6. The four perspectives of the *Balanced Scorecard* model (Kaplan and Norton 1992). With permission

The concept of *balance* consists in placing the right emphasis on all processes/organizations' important *dimensions*. In fact, usually performance indicators tend to consider only the economic/financial dimension (see Sect. 1.4.6).

5.4.3 The "Critical Few" method

Managing too many indicators may produce different drawbacks: (1) losing sight of all the indicators' impact, (2) distracting management's focus from those indicators that are the most critical to organizational success, (3) not identifying the correlation/influence between two indicators, and so on. Consequently, the goal is to reduce the number of indicators as much as possible: *pluritas non est potenda sine necessitate* (Thorborn 1918). The process of simplifying and distilling a large number of performance measures across the organization to select a "critical few" should be viewed as part of the performance measurement process itself. It helps sharpen understanding of the strategic plan and its supporting objectives.

5.5   The problem of indicators' synthesis

The problem of indicators' synthesis has been introduced in Sect. 5.4, when presenting the performance dashboard. The dashboard is characterized by a small number of indicators, providing a compressed vision of the system's global performances. The present section will argue how these indicators can be identified.

Synthesis indicators can be selected according to different operational approaches. In the following sub-sections we provide a detailed description of three of them:

a. synthesis based on the concept of *"relative importance"*;
b. "minimum set covering" synthesis;
c. synthesis based on the indicators degree of correlation.

Approach (a) provides a synthesis based on the representation-targets' importance, and the "relations" among indicators and representation-targets. The aim of this technique is to extract a subset of indicators focussing on a few important aspects. So, the resulting performance dashboard does not consider all the representation-targets, but only the most important ones.

Approach (b) selects the smallest indicators set, which influences all the process targets.

Thirdly, approach (c) is based on the indicators degree of correlation. The concept of degree of correlation is more extensive than the concept of statistical correlation. Correlation degree is expressed in *qualitative* terms, and indicates an indicator's influence on other indicators (and the *vice versa*) (Franceschini 2002). For example, the indicator 3.1 "training of the personnel" – presented in Sect. 5.4.3 – may influence both indicator 3.3 "customer's perception of the operator's competence" and indicator 6.1 "response clarity". In a Help Desk, increasing the personnel's training enable employees to be more competent and – consequently – responses are clearer. This correlation is qualitative and is based on empirical considerations.

It is important to note that, at the present time, an "optimal" synthesis technique is still missing. The choice of a specific technique depends on the examined process, the type of representation, and the characteristics of the available data.

5.5.1 Indicators synthesis based on the concept of "relative importance"

By using the information contained in the Relationship Matrix (see Fig. 5.1 and Fig. 5.2) and using as starting point the importance assigned to the representation-targets, we can establish a ranking for the indicators.

The classic method used for ranking the indicators is the *Independent Scoring Method* (Akao 1998; Franceschini 2002). It requires two operative steps. The first one consists in converting the symbolic relationships between representation-targets and indicators into "equivalent" numerical values. This conversion from an ordinal to a cardinal scale utilizes 1-3-9, or 1-3-5, or 1-5-9 scales. If we use only one symbol, it can be converted into value 1. This sort of conversion is very delicate. It hides the change from an ordinal scale to a cardinal (Franceschini and Romano 1999; Franceschini 2002).

The second step entails determining the level of importance w_j of each indicator. It is obtained by summing up the products of relative importance for each representation-target, multiplied by the *quantified* value of the relationship existing between that j-th indicator and each of the representation-targets related to it. We obtain:

$$w_j = \sum_{i=1}^{m} d_i \cdot r_{ij} \tag{5.1}$$

where:

d_i is the degree of relative importance of the i-th representation-target, $i = 1, 2\ldots m$;

r_{ij} is the cardinal relationship between the i-th representation-target and the j-th indicator, $i = 1, 2\ldots m, j = 1, 2, \ldots, n$;

w_j is the technical importance rating of the j-th indicator, $j = 1, 2\ldots n$;

m is the number of representation-targets;

n is the number of indicators.

The (absolute) importance level (w_j) of indicator j may be transformed into a relative importance level (w^*_j), expressed as a percentage:

$$w^*_j = \frac{w_j}{\sum_{j=1}^{n} w_j} \tag{5.2}$$

where $j = 1, 2 \ldots n$ is the indicators number.

These values (w_j and w^{*}_{j}) are included in the Relationship Matrix, classified as *absolute importance* and relative importance respectively (see Fig. 5.9).

Returning to the example presented in Sect. 5.3.1, symbols are converted into numerical values according to the following encoding:

$$● = 9; \qquad O = 3; \qquad \Delta = 1 \qquad (5.3)$$

By applying the *Independent Scoring Method*, the absolute importance of indicators is calculated (see Fig. 5.9).

		Performance measurements								
		Routing effectiveness	Reply accuracy	Uniformity of replies (from different operators)	Time for requests implementation	Competence perception	Answered calls percentage	Courtesy of responses	Security of data keeping	Number of active lines
Targets	**Importance**									
Reliability	5	●	O	●	O	Δ			O	
Responsiveness	5					●	Δ			
Competence	4	●	●	●		●				
Access	4						●			O
Courtesy	3			Δ				●		
Communication	3		Δ	●		Δ		O		
Credibility	3	Δ			O	O				
Security	5	Δ							●	
Understanding/ Knowing the customer	4	O	Δ	Δ		Δ		O		
Tangibles	3									●
Absolute importance		101	58	124	60	62	36	48	60	39
Relative importance (%)		17.18	9.86	21.09	10.20	10.54	6.12	8.16	10.20	6.63
Present model values		93%	A	A	22 min	MB	98%	M	3%	3
Target values		>90%	A	A	20 min	MB	>99%	A	<5%	5

Legend:
Δ=1 (weak relationship)
O=3 (medium relationship)
●=9 (strong relationship)

Fig. 5.9. Calculation of the indicators' importance with reference to the Relationship Matrix for a Help Desk service. The relations between "targets" and "performance measurement" are coded by specific symbols: a triangle for weak relationships (numerical coding: $\Delta = 1$), a circle for medium (numerical coding: O = 3) and two concentric circles for strong (numerical coding: ● = 9) (Franceschini 2002).

The selection of a critical few set of performance is carried out considering those indicators with the highest (relative) importance value. In the example of Fig. 5.9, defining a 10% "cut threshold" (indicators with relative importance lower than 10% are not included in the *Critical Few* set), we obtain the following indicators set:

- uniformity of replies from different operators (21.09 %);
- routing effectiveness (17.18 %);
- competence perception (10.54 %);
- security of data keeping (10.20 %);
- time for requests implementation (10.20 %).

The "cut threshold" value depends on the examined process peculiarities. For example the expected number of indicators making up the *Critical Few* set, or the established minimum value of relative importance. Evidently, the higher the cut threshold, the lower the number of selected indicators.

The presented distillation method, in spite of providing a selection of the most important indicators, does not guarantee a complete coverage. Furthermore, this technique does not take into account of possible correlations among indicators.

To guarantee a concurrent monitoring of all the representation-targets, it is necessary to use other selection methods for indicators, as shown in the next section.

5.5.2 Indicators synthesis based on the concept of "minimum set covering"

In some situations – rather than selecting the most important indicators – it is more important to define the *minimum set* of indicators able to cover all the representation-targets. This method can be useful when we want to establish a minimum set of indicators providing a global vision of the process condition. Obviously, this does not mean that some indicators can be neglected during the process monitoring. It does imply that process can be organized in such a way to give more importance to those indicators that better cover the representation-targets.

The search for the minimum of indicators covering all representation-targets is a classic combinatorial optimization problem, known as the *set covering problem* (Nemhauser and Wolsey 1988; Parker and Rardin 1988).

In more detail, if $M = \{1,...,m\}$ is a finite set and $\{M_j\}$, for $j \in N = \{1,...,n\}$, a given collection of subsets of M, we say that $F \subseteq N$ covers M if $\bigcup_{j \in F} M_j = M$.

The sets M_j are known as *covering sets*. If c_j is the cost (weight) associated to each M_j, the *minimum set covering* problem becomes that of *minimum-cost set covering*.

The search for the minimum number of columns (indicators) able to cover all rows (representation-targets) is a *set covering* problem with $c_j = 1, \forall j \in N$. The set covering problem has a nonpolynominial computational complexity, which increases together with the problem dimension (Parker and Rardin 1988).

In this specific case, the aim is to give an *agile* supporting tool to the process monitoring system designer, so can use a heuristic algorithm which has a polynominial computational complexity. The algorithm is particularly suitable to give quick responses in a short time. This algorithm is known as Nemhauser's (Nemhauser and Wolsey 1988).

Nemhauser's algorithm

Initialization:
  $M^1 = M$, $N^1 = N$, $t = 1$
  Solution generation at step $t > 1$:
  (a) calculate c_j;   select $j' \in N : c_t = \min_j \{c_j / \max |M_j \cap M'|\}$;
  $N^{t+1} = N^t \setminus \{j'\}$; $M^{t+1} = M^t \setminus M_j$
  If $M^{t+1} \neq \varnothing$ then $t = t+1$, go to *step (a)*
  If $M^{t+1} = \varnothing$ then *stop*
  The final solution is given by all elements $j \notin N^{t+1}$.

The following steps describe the algorithm *modus operandi*:

1. Select the indicator which has the maximum number of relations with representation-targets. The relation intensity (weak, medium, strong) is not considered. In case of equivalent indicators, the indicator with low cost associated is selected. In case of further equivalence, selection is indifferent;
2. The selected indicator is removed from the Relationship Matrix, and it is included in the *Critical Few* set;

3. For the remaining Relationship Matrix indicators, we eliminate the symbols linked to the representation-targets covered by the indicator selected at step 2;
4. The procedure is repeated until all the Relationship Matrix symbols are removed.

Fig. 5.10 returns to the *Help Desk* example (Sect. 5.3.1).

In first step, two possible indicators can be selected from Relationship ("Uniformity of replies from different operators" and "Competence perception"). For simplicity, and assuming the indicators' cost the same, we decide to select the first one, including it in the *Critical Few* set.

		Performance measurements								
Targets	**Importance**	Routing effectiveness	Reply accuracy	Uniformity of replies (from different operators)	Time for requests implementation	Competence perception	Answered calls percentage	Courtesy of responses	Security of data keeping	Number of active lines
Reliability	5	●	O	●	O	Δ			O	
Responsiveness	5				●	Δ				
Competence	4	●	●	●		●				
Access	4						●			O
Courtesy	3			Δ				●		
Communication	3		Δ	●		Δ		O		
Credibility	3	Δ		O		O				
Security	5	Δ							●	
Understanding/ Knowing the customer	4	O	Δ	Δ		Δ		O		
Tangibles	3									●

Fig. 5.10. Application of Nemhauser's algorithm. The first step consists in identifying the indicator which has the maximum number of relations with representation-targets. In this example, we assume the indicators' cost the same

Then, we eliminate all the Relationship Matrix symbols placed in the rows covered by selected indicator (rows 1, 3, 5, 6, 7, 9), with reference to the remaining indicators (see Fig. 5.11). Considering the new Relationship Matrix, the procedure is repeated. The next selected indicator is "Number of active lines". New Relationship Matrix is shown in Fig. 5.12.

Among four possible indicators in the Relationship Matrix ("Routing effectiveness", "Time for requests implementation", "Competence perception" and "Security of data keeping" in Fig. 5.12), we select the first one – "Routing effectiveness" – and we include it in the *Critical Few* set.

Fig. 5.13 shows the new changes.

		Performance measurements								
Targets	Importance	Routing effectiveness	Reply accuracy	Uniformity of replies (from different operators)	Time for requests implementation	Competence perception	Answered calls percentage	Courtesy of responses	Security of data keeping	Number of active lines
Reliability	5									
Responsiveness	5					●	Δ			
Competence	4									
Access	4						●			O
Courtesy	3									
Communication	3									
Credibility	3									
Security	5	Δ							●	
Understanding/ Knowing the customer	4									
Tangibles	3									●

Fig. 5.11. Application of Nemhauser's. The second step consists in identifying – among the remaining indicators – the indicator which covers the maximum number of representation-targets (shown in light grey)

		Performance measurements								
Targets	Importance	Routing effectiveness	Reply accuracy	Uniformity of replies (from different operators)	Time for requests implementation	Competence perception	Answered calls percentage	Courtesy of responses	Security of data keeping	Number of active lines
Reliability	5									
Responsiveness	5					●	Δ			
Competence	4									
Access	4									
Courtesy	3									
Communication	3									
Credibility	3									
Security	5	Δ							●	
Understanding/ Knowing the customer	4									
Tangibles	3									

Fig. 5.12. Application of Nemhauser's algorithm with reference to the Relationship Matrix for a Help Desk service. The third step consists in identifying – among the remaining indicators – the indicator which covers the maximum number of representation-targets (shown in light grey)

Targets	Importance	Routing effectiveness	Reply accuracy	Uniformity of replies (from different operators)	Time for requests implementation	Competence perception	Answered calls percentage	Courtesy of responses	Security of data keeping	Number of active lines
		Performance measurements								
Reliability	5									
Responsiveness	5				●	Δ				
Competence	4									
Access	4									
Courtesy	3									
Communication	3									
Credibility	3									
Security	5									
Understanding/ Knowing the customer	4									
Tangibles	3									

Fig. 5.13. Application of Nemhauser's algorithm with reference to the Relationship Matrix for a Help Desk service. The fourth step consists in identifying – among the remaining indicators – the indicator which covers the maximum number of representation-targets (shown in light grey). The dark grey columns refer to the indicators already included in the *Critical Few* set

Finally, between the two possible indicators ("Time for requests implementation" and "Competence perception"), we select the first one.

In conclusion, the *Critical Few* indicators set is given by:

- "Uniformity of replies from different operators";
- "Number of active lines";
- "Routing effectiveness";
- "Time for requests implementation".

The suggested algorithm does not consider either the representation-target importance or the relation intensity (weak, medium, strong) between indicators and representation-targets. This information can be used: (1) translating the indicators' importance value (calculated applying the *Independent Scoring Method*) into weight (c_j); (2) considering the sum of the quantified values of the relationship existing between that j-th indicator and each of the representation-targets related to it:

$$c_j = \sum_{i=1}^{m} r_{ij} \qquad (5.4)$$

r_{ij} is the cardinal relationship between the i-th representation-target and the j-th indicator, $i = 1, 2 \ldots m, j = 1, 2 \ldots n$.

Considering this enhancement, the algorithm changes as follows.

Modified Nemhauser's algorithm

Initialization:

$M^1 = M$, $N^1 = N$, $t = 1$

Solution generation at step $t > 1$:

calculate c_j; select $j' \in N : c_t = \max_j \left\{ c_j / \max \left| M_j \cap M' \right| \right\}$;

$N^{t+1} = N^t \setminus \{ j' \}$; $M^{t+1} = M^t \setminus M_j$

If $M^{t+1} \neq \emptyset$ then $t = t + 1$, go to *step (a)*

If $M^{t+1} = \emptyset$ then *stop*

The final solution is given by all elements $j \notin N^{t+1}$.

The algorithm logic is based on the following steps:

1. Selecting the indicator which has the maximum number of relations with representation-targets. The relation intensity (weak, medium, strong) is not considered. In the case of equivalent indicators, the indicator with the highest weight (c_j) is selected.
2. The selected indicator is removed from the Relationship Matrix, and it is included in the *Critical Few* set.
3. For the Relationship Matrix remaining indicators, we eliminate the symbols linked to the representation-targets covered by the indicator selected at step 2.
4. Indicators weights (c_j) are re-calculated, only using the remaining coefficients (the quantified relationships between indicators and re-presentation-targets).
5. The procedure is repeated until all the Relationship Matrix symbols are removed.

This new algorithm is now applied to the *Help Desk* example (Sect. 5.3.1). Each indicator is related to a cost (weight) c_j, calculated applying the *Independent Scoring Method*, with the same encoding of Eq. 5.3 (see Fig. 5.14).

In first step, the selected indicator is "Uniformity of replies from different operators" ($c_3 = 21.09\%$), which has the same number of relations as the indicator "Competence perception". Again, for simplicity, assuming indicators' weights are the same, we decide to select the first one, including it in the *Critical Few* set.

		Performance measurements								
Δ=1 (weak relationship) O=3 (medium relationship) ●=9 (strong relationship) **Targets**	**Importance**	Routing effectiveness	Reply accuracy	Uniformity of replies (from different operators)	Time for requests implementation	Competence perception	Answered calls percentage	Courtesy of responses	Security of data keeping	Number of active lines
Reliability	5	●	O	●	O	Δ			O	
Responsiveness	5				●	Δ				
Competence	4	●	●	●		●				
Access	4						●			O
Courtesy	3			Δ				●		
Communication	3		Δ	●		Δ		O		
Credibility	3	Δ		O		O				
Security	5	Δ							●	
Understanding/ Knowing the customer	4	O	Δ	Δ		Δ			O	
Tangibles	3									●

Absolute importance	101	58	124	60	62	36	48	60	39
Relative importance (%)	17.18	9.86	21.09	10.20	10.54	6.12	8.16	10.20	6.63

Fig. 5.14 Application of modified Nemhauser's algorithm with reference to the Relationship Matrix for a Help Desk service. The first step consists in identifying (in light grey) the indicator which has the maximum number of relations with representation-targets (independently on the relation intensity). Indicators' weight coefficients (c_j) are calculated applying the *Independent Scoring Method*. In the case of equivalent indicators, the indicator with the highest weight is selected

Second step consists in removing all the symbols in the Relationship Matrix rows "covered" by the selected indicator. Then, the remaining indicators weights (c_j) are re-calculated. The new selected indicator is "Number of active lines" (Fig. 5.15). Fig. 5.16 shows the new Relationship Matrix.

Each of the four remaining indicators (in light grey in Fig. 5.16) has a single relationship with representation-targets. The relative weight of the indicators "Time for requests implementation" and "Security of data keeping", is the same (45%). The first one is selected.

		Performance measurements								
Δ=1 (weak relationship) O=3 (medium relationship) ●=9 (strong relationship)		Routing effectiveness	Reply accuracy	Uniformity of replies (from different operators)	Time for requests implementation	Competence perception	Answered calls percentage	Courtesy of responses	Security of data keeping	Number of active lines
Targets	Importance									
Reliability	5									
Responsiveness	5					●	Δ			
Competence	4									
Access	4						●			O
Courtesy	3									
Communication	3									
Credibility	3									
Security	5	Δ							●	
Understanding/ Knowing the customer	4									
Tangibles	3									●
Absolute importance		5	0	0	45	5	36	0	45	39
Relative importance (%)		2.86	0.00	0.00	25.71	2.86	20.57	0.00	25.71	22.29

Fig. 5.15. Application of modified Nemhauser's. The second step consists in identifying – among the remaining indicators – the indicator with the maximum number of relations with the representation-targets (in light grey). In the case of equivalent indicators, the indicator with the highest weight (c_j) is selected. The dark grey columns refer to the indicators included in the *Critical Few* set

		Performance measurements								
Δ=1 (weak relationship) O=3 (medium relationship) ●=9 (strong relationship)		Routing effectiveness	Reply accuracy	Uniformity of replies (from different operators)	Time for requests implementation	Competence perception	Answered calls percentage	Courtesy of responses	Security of data keeping	Number of active lines
Targets	Importance									
Reliability	5									
Responsiveness	5					●	Δ			
Competence	4									
Access	4									
Courtesy	3									
Communication	3									
Credibility	3									
Security	5	Δ							●	
Understanding/ Knowing the customer	4									
Tangibles	3									
Absolute importance		5	0	0	45	5	0	0	45	0
Relative importance (%)		5.00	0.00	0.00	45.00	5.00	0.00	0.00	45.00	0.00

Fig. 5.16. Application of modified Nemhauser's algorithm. The third step consists in identifying – among the remaining indicators – the indicator with the maximum number of relations with representation-targets (in light grey). In the case of equivalent indicators, the indicator with the highest weight (c_j) is selected

Fig. 5.17 shows the new Relationship Matrix.

Finally, between the two remaining indicators we select the indictor "Security of data keeping" ($c_8 = 90\%$).

		Performance measurements								
Δ=1 (weak relationship)		Routing effectiveness	Reply accuracy	Uniformity of replies (from different operators)	Time for requests implementation	Competence perception	Answered calls percentage	Courtesy of responses	Security of data keeping	Number of active lines
Ο=3 (medium relationship)										
●=9 (strong relationship)										
Targets	**Importance**									
Reliability	5									
Responsiveness	5									
Competence	4									
Access	4									
Courtesy	3									
Communication	3									
Credibility	3									
Security	5	Δ							●	
Understanding/ Knowing the customer	4									
Tangibles	3									
Absolute importance		5	0	0	0	5	0	0	45	0
Relative importance (%)		10.00	0.00	0.00	0.00	5.00	0.00	0.00	90.00	0.00

Fig. 5.17. Application of modified Nemhauser's algorithm with reference to the Relationship Matrix for a Help Desk service. The fourth step consists in identifying – among the remaining indicators – the indicator with the maximum number of relations with representation-targets (in light grey). In the case of equivalent indicators, the indicator with the highest weight (c_j) is selected

In this case, the *Critical Few* indicators set is given by:

- "Uniformity of replies from different operators" (6 relationships with the "uncovered" representation-targets, and $c_3 = 21.09\%$);
- "Number of active lines" (2 relationships with the "uncovered" representation-targets, and $c_9 = 22.29\%$);
- "Time for requests implementation" (1 relationship with the "uncovered" representation-targets, and $c_4 = 45\%$);
- "Security of data keeping" (1 relationship with the "uncovered" representation-targets, and $c_8 = 90\%$).

The same example is now presented calculating weights (c_j) as the sum of the relationships' quantified values.

		Performance measurements								
Δ=1 (weak relationship) O=3 (medium relationship) ●=9 (strong relationship) **Targets**	**Importance**	Routing effectiveness	Reply accuracy	Uniformity of replies (from different operators)	Time for requests implementation	Competence perception	Answered calls percentage	Courtesy of responses	Security of data keeping	Number of active lines
Reliability	5	●	O	●	O	Δ			O	
Responsiveness	5				●	Δ				
Competence	4	●	●	●		●				
Access	4						●			O
Courtesy	3			Δ				●		
Communication	3		Δ	●		Δ		O		
Credibility	3	Δ		O		O				
Security	5	Δ							●	
Understanding/ Knowing the customer	4	O	Δ	Δ		Δ		O		
Tangibles	3									●
Sum of weights (c_j)		23	14	32	12	16	9	15	12	12

Fig. 5.18. Application of modified Nemhauser's algorithm. The first step consists in identifying (in light grey) the indicator with the maximum number of relations with representation-targets. Indicators' weight coefficients (c_j) are calculated as the sum of the relationships' quantified values. In the case of equivalent indicators, the indicator with the highest weight is selected

		Performance measurements								
Δ=1 (weak relationship) O=3 (medium relationship) ●=9 (strong relationship) **Targets**	**Importance**	Routing effectiveness	Reply accuracy	Uniformity of replies (from different operators)	Time for requests implementation	Competence perception	Answered calls percentage	Courtesy of responses	Security of data keeping	Number of active lines
Reliability	5									
Responsiveness	5				●	Δ				
Competence	4									
Access	4						●			O
Courtesy	3									
Communication	3									
Credibility	3									
Security	5	Δ							●	
Understanding/ Knowing the customer	4									
Tangibles	3									●
Sum of weights (c_j)		1	0	0	9	1	9	0	9	12

Fig. 5.19. Application of modified Nemhauser's algorithm. The second step consists in identifying – among the remaining indicators – the indicator with the maximum number of relations with the representation-targets (shown in light grey. Dark grey columns refer to the indicators already included in the *Critical Few* set

						Performance measurements				
Δ=1 (weak relationship) O=3 (medium relationship) ●=9 (strong relationship)		Routing effectiveness	Reply accuracy	Uniformity of replies (from different operators)	Time for requests implementation	Competence perception	Answered calls percentage	Courtesy of responses	Security of data keeping	Number of active lines
Targets	**Importance**									
Reliability	5									
Responsiveness	5					O	Δ			
Competence	4									
Access	4									
Courtesy	3									
Communication	3									
Credibility	3									
Security	5	Δ							●	
Understanding/ Knowing the customer	4									
Tangibles	3									
Sum of weights (c_j)		1	0	0	9	1	0	0	9	0

Fig. 5.20. Application of modified Nemhauser's algorithm. The third step consists in identifying – among the remaining indicators – the indicator with the maximum number of relations with representation-targets (in light grey). In the case of equivalent indicators, the indicator with the highest weight (c_j) is selected. The dark grey columns refer to the indicators included in the *Critical Few* set

						Performance measurements				
Δ=1 (weak relationship) O=3 (medium relationship) ●=9 (strong relationship)		Routing effectiveness	Reply accuracy	Uniformity of replies (from different operators)	Time for requests implementation	Competence perception	Answered calls percentage	Courtesy of responses	Security of data keeping	Number of active lines
Targets	**Importance**									
Reliability	5									
Responsiveness	5									
Competence	4									
Access	4									
Courtesy	3									
Communication	3									
Credibility	3									
Security	5	Δ							●	
Understanding/ Knowing the customer	4									
Tangibles	3									
Sum of weights (c_j)		1	0	0	0	0	0	0	9	0

Fig. 5.21. Application of modified Nemhauser's algorithm. The fourth step consists in identifying – among the remaining indicators – the indicator with the maximum number of relations with representation-targets (in light grey). The dark grey columns refer to the indicators already included in the *Critical Few* set

The *Critical Few* indicators set is given by:

- "Uniformity of replies from different operators": 6 relationships with all the representation-targets, and $c_8 = 32$ (Fig. 5.18);
- "Number of active lines": 2 relationships with the remaining "uncovered" representation-targets, and $c_9 = 12$ (Fig. 5.19);
- "Time for requests implementation": 1 relationship with the remaining "uncovered" representation-targets, and $c_4 = 9$ (Fig. 5.20);
- "Security of data keeping": 1 relationship with the remaining "uncovered" representation-targets, and $c_8 = 9$ (Fig. 5.21).

We notice that the results obtained in the two previous examples are identical. In general, these two methods lead to different *Critical Few* sets.

We remark that the *Independent Scoring method* considers both the representation-targets importance, and the relationships intensity (weak, medium, strong). The second technique only considers the relationships' intensity, neglecting the representation targets' importance.

Nemhauser's algorithm, in both these versions, guarantees a complete "covering" of the representation-targets, but does not provide a minimization of the selected indicators set. Furthermore, it ignores the correlations among indicators. Ideally, *Critical Few* indicators should guarantee a complete representation-target covering, but should also have a low correlation. The method presented in the next section tries to fulfil both these requirements.

5.5.3 Indicators synthesis based on the concept of "degree of correlation"

Using QFD terminology, two indicators are defined as correlated if variations on the first one determine variations on the second and *vice versa* (Franceschini 2002). At present, the process analysts establish correlations among indicators on the basis of merely *qualitative* reasoning. By observing a general Relationship Matrix, it may be noted that in many cases correlated indicators influence the same representation-targets. This can be a starting point in building a partially automatic tool to indirectly define correlations among indicators. As a matter of fact, if the *i*-th indicator influences some precise representation-targets, it is likely that the *j*-th indicator correlated to it influences the same representation-targets. Moreover, if the *dependence* among indicators *induced* by the *action* of the same indicators may imply the presence of a correlation, the opposite it is not necessarily true. In fact, it can be demonstrated that a correlation between two indicators may exist, without *induced links* on representation-targets in the Relationship Matrix (Franceschini 2002). Consequently, the method proposed here investigates the *induced dependence* and can highlight only a fraction of the total correlations. The presence of an induced dependence on the indicators is, therefore, a necessary but not a sufficient condition to state that two indicators are correlated. It is the designer, playing a new role of "validator", who must confirm the possible sufficiency.

To formulate the existence of dependencies induced by indicators, an n-dimensional space constituted by a set of column vectors $\mathbf{b}_j \in \mathfrak{R}^n$ is considered (each one associated to a well-defined indicator in the Relationship Matrix). Supposing that the Relationship Matrix $\mathbf{R}$ is filled adopting the symbol $\mathbf{O}$ to individuate strong relationships, the symbol O for medium relationships, and the symbol Δ for weak relationships, the coefficients of vectors $\mathbf{b}_j$ ($\forall j = 1,...,n$) are determined as the following:

$\forall i, j$

if $r_{i,j} = \mathbf{O}$ or $r_{i,j} = O$ or $r_{i,j} = \Delta$;

then $b_{i,j} = 1$;

otherwise $b_{i,j} = 0$.

As an alternative, to differentiate the relations intensity (strong, medium, weak), the standard encoding ($\mathbf{O} = 9$, $O = 3$, $\Delta = 1$) can be used.

Thus, by starting from the symbolic matrix $\mathbf{R}$ a new binary matrix $\mathbf{B} \in \mathfrak{R}^{m,n}$ is created. Matrix $\mathbf{B}$ columns ($\mathbf{b}_j$) are then normalized producing another matrix $\mathbf{N} \in \mathfrak{R}^{m,n}$, with the columns named $\mathbf{v}_j$ ($\forall j = 1,...,n$).

Vector $\mathbf{v}_j$ components are obtained as follows:

$$v_{i,j} = \frac{b_{i,j}}{\sqrt{\sum_{i=1}^{m} b_{i,j}^2}} \qquad \forall i, j \tag{5.5}$$

The examples in Fig. 5.22 and Fig. 5.23 can give a better idea of the building process of matrix $\mathbf{N}$.

To represent the effects of the interdependence between i-th and j-th indicators, the coefficient $q_{i,j}$ (scalar products of vectors $\mathbf{v}_i$) is introduced:

$$q_{i,j} = \mathbf{v}_i \cdot \mathbf{v}_j = \cos\left(\mathbf{v}_i, \mathbf{v}_j\right) \qquad \forall i, j = 1,...,n \tag{5.6}$$

$$R = \begin{array}{|c|c|c|}\hline \bullet & O & \\\hline \Delta & & O \\\hline & \Delta & \\\hline\end{array} \quad B = \begin{array}{|c|c|c|}\hline 1 & 1 & 0 \\\hline 1 & 0 & 1 \\\hline 0 & 1 & 0 \\\hline\end{array} \quad N = \begin{array}{|c|c|c|}\hline 1/\sqrt{2} & & 0 \\\hline 1/\sqrt{2} & 0 & 1 \\\hline 0 & 1/\sqrt{2} & 0 \\\hline\end{array}$$

Fig. 5.22. Example of the building process of matrix **N** starting from matrix **R** (symbols are codified using a binary encoding: ●, O, Δ = 1)

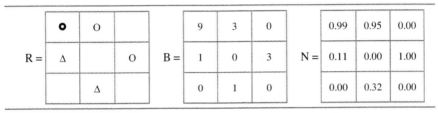

$$R = \begin{array}{|c|c|c|}\hline \bullet & O & \\\hline \Delta & & O \\\hline & \Delta & \\\hline\end{array} \quad B = \begin{array}{|c|c|c|}\hline 9 & 3 & 0 \\\hline 1 & 0 & 3 \\\hline 0 & 1 & 0 \\\hline\end{array} \quad N = \begin{array}{|c|c|c|}\hline 0.99 & 0.95 & 0.00 \\\hline 0.11 & 0.00 & 1.00 \\\hline 0.00 & 0.32 & 0.00 \\\hline\end{array}$$

Fig. 5.23. Example of the building process of matrix **N** starting from matrix **R** (symbols are codified using a standard encoding: ● = 9, O = 3, Δ = 1)

By calculating $q_{i,j}$ for all pairs of vectors in the **N** matrix, it is possible to determine the indicators correlation matrix **Q**:

$$Q = N^T N \qquad (5.7)$$

$Q \in \Re^{m,n}$ is symmetrical, with $q_{i,i} = 1$, $\forall i = 1,...,n$.

Matrix **Q** expresses the degree of induced dependence among indicators with reference to their capacity of influencing the same representation-targets. It may be observed that, if one works with large matrices, the determination of matrix **Q** reveals the existence of columns of rows without relation with other columns or rows of the matrix **R**, respectively. This fact is highlighted by appearance of some zeros in the main diagonal of **Q**.

Information contained in **Q** are compared with a prefixed threshold k (with $0 \le k \le 1$); $\forall i, j$, if $q_{i,j} > k$ then a *potential* correlation between indicators i-th and j-th is admitted, else this correlation is supposed nonexistent. So starting from **Q**, a new matrix $\hat{Q}$ is built.

Considering the *Call Center* example (Fig. 5.9) and using a standard symbols encoding (● = 9; O = 3; Δ = 1), we obtain matrices **R**, **B**, **N** and **Q** respectively shown in Table 5.12, 5.13, 5.14 and 5.15.

Fixing the threshold value as $k = 0.6$, we obtain the correlation matrix $\hat{Q}$ in Fig. 5.24.

Table 5.12. Matrix of the relations (**R**) among representation-targets and indicators (see Fig. 5.9)

$R =$

●	O	●	O	Δ			O	
			●	Δ				
●	●	●		●				
					●			O
		Δ				●		
	Δ	●		Δ		O		
Δ		O		O				
Δ							●	
O	Δ	Δ		Δ		O		
								●

Table 5.13. Matrix of the encoded relations (**B**) (see Table 5.12). Standard encoding: Δ = 1; O = 3; ● = 9

$B =$

9	3	9	3	1			3	
			9	1				
9	9	9		9				
					9			3
		1				9		
	1	9		1		3		
1		3		3				
1							9	
3	1	1		1		3		
								9

Table 5.14. Matrix of the normalized relations (**N**) (see Table 5.13)

$N =$

0.68	0.31	0.56	0.32	0.10	0.00	0.00	0.32	0.00
0.00	0.00	0.00	0.95	0.10	0.00	0.00	0.00	0.00
0.68	0.94	0.56	0.00	0.93	0.00	0.00	0.00	0.00
0.00	0.00	0.00	0.00	0.00	1.00	0.00	0.00	0.32
0.00	0.00	0.06	0.00	0.00	0.00	0.90	0.00	0.00
0.00	0.10	0.56	0.00	0.10	0.00	0.30	0.00	0.00
0.08	0.00	0.19	0.00	0.31	0.00	0.00	0.00	0.00
0.08	0.00	0.00	0.00	0.00	0.00	0.00	0.95	0.00
0.23	0.10	0.06	0.00	0.10	0.00	0.30	0.00	0.00
0.00	0.00	0.00	0.00	0.00	0.00	0.00	0.00	0.95

Starting from matrix $\hat{Q}$, *Critical Few* indicators set can be selected according to the following steps:

1. Selecting the indicator which has the maximum number of correlations with other ones (in case of equivalent indicators and if indicators costs

are known, we select the lowest cost indicator. In case of further equality, the choice is indifferent);

2. The selected indicator is included in the *Critical Few* set, and the correlations related to it are removed from matrix $\hat{Q}$;
3. The procedure is repeated until all the Relationship Matrix correlations are removed (except those in the matrix main diagonal);
4. The remaining indicators are included in the *Critical Few* set.

Table 5.15. Matrix of the correlations (**Q**) among indicators for the example of Fig. 5.9

1.00	0.88	0.80	0.22	0.75	0.00	0.07	0.29	0.00
0.88	1.00	0.77	0.10	0.92	0.00	0.06	0.10	0.00
0.80	0.77	1.00	0.18	0.71	0.00	0.25	0.18	0.00
0.22	0.10	0.18	1.00	0.13	0.00	0.00	0.10	0.00
0.75	0.92	0.71	0.13	1.00	0.00	0.06	0.03	0.00
0.00	0.00	0.00	0.00	0.00	1.00	0.00	0.00	0.32
0.07	0.06	0.25	0.00	0.06	0.00	1.00	0.00	0.00
0.29	0.10	0.18	0.10	0.03	0.00	0.00	1.00	0.00
0.00	0.00	0.00	0.00	0.00	0.32	0.00	0.00	1.00

$\mathbf{Q} =$ (to the left of the matrix)

	(I_1) Routing effectiveness	(I_2) Reply accuracy	(I_3) Uniformity of replies (different operators)	(I_4) Time for requests implementation	(I_5) Competence perception	(I_6) Answered calls percentage	(I_7) Courtesy of responses	(I_8) Security of data keeping	(I_9) Number of active lines
(I_1) Routing effectiveness	X	X	X		X				
(I_2) Reply accuracy	X	X	X		X				
(I_3) Uniformity of replies (from different operators)	X	X	X						
(I_4) Time for requests implementation				X					
(I_5) Competence perception	X	X			X				
(I_6) Answered calls percentage						X			
(I_7) Courtesy of responses							X		
(I_8) Security of data keeping								X	
(I_9) Number of active lines									X

Fig. 5.24 Matrix of the correlations ($\hat{Q}$) among indicators for the example of Fig. 5.9. In this specific case threshold value (*k*) is 0.75. Correlations are identified by the symbol X

The application of the procedure to the indicators in Fig. 5.24 produces the following *Critical Few* set:

- "Routing effectiveness", in the first step (Fig. 5.25);
- "Time for requests implementation", "Answered calls percentage", "Courtesy of responses", "Security of data keeping" and "Number of active lines", in the second step (Fig. 5.26).

	(I_1) Routing effectiveness	(I_2) Reply accuracy	(I_3) Uniformity of replies (different operators)	(I_4) Time for requests implementation	(I_5) Competence perception	(I_6) Answered calls percentage	(I_7) Courtesy of responses	(I_8) Security of data keeping	(I_9) Number of active lines
(I_1) Routing effectiveness	X	X	X		X				
(I_2) Reply accuracy	X	X	X		X				
(I_3) Uniformity of replies (from different operators)	X	X	X						
(I_4) Time for requests implementation				X					
(I_5) Competence perception	X	X			X				
(I_6) Answered calls percentage						X			
(I_7) Courtesy of responses							X		
(I_8) Security of data keeping								X	
(I_9) Number of active lines									X

Fig. 5.25. Matrix of the correlations ($\hat{Q}$) among indicators in Fig. 5.9. Threshold value (k) is 0.75. Correlations are identified by the symbol X. Indicators with the highest number of correlations are highlighted in light grey. The first indicator included in the *Critical Few* set is selected from these

In general, this approach does not guarantee a "complete covering" of the representation-targets. The covering level depends on the prefixed threshold value (k). Here follows a heuristic technique to guarantee a complete covering of all representation-targets:

1. Matrix $\hat{Q}$ is constructed using a prefixed value of k.
2. Indicators which are not correlated to each other are included in the *Critical Few* set.
3. The indicator with the maximum number of correlations is selected. In the case of equivalent indicators and if indicators costs are known, the lowest cost indicator is selected. In case of further equality, the choice is indifferent. The selected indicator is included in the Critical Few set.

4. Correlations related to selected indicator are removed from matrix $\hat{Q}$ (except those in the matrix main diagonal).

5. The procedure is repeated until matrix $\hat{Q}$ is empty (except those in the matrix main diagonal);a check of selected indicators is done to assess whether they guarantee a complete covering. If so, the procedure stops. Otherwise threshold k is increased, and the procedure restarts from point (1).

	(I_1) Routing effectiveness	(I_2) Reply accuracy	(I_3) Uniformity of replies (different operators)	(I_4) Time for requests implementation	(I_5) Competence perception	(I_6) Answered calls percentage	(I_7) Courtesy of responses	(I_8) Security of data keeping	(I_9) Number of active lines
(I_1) Routing effectiveness	X	X	X		X				
(I_2) Reply accuracy	X	X	X		X				
(I_3) Uniformity of replies (from different operators)	X	X	X						
(I_4) Time for requests implementation				X					
(I_5) Competence perception	X	X			X				
(I_6) Answered calls percentage						X			
(I_7) Courtesy of responses							X		
(I_8) Security of data keeping								X	
(I_9) Number of active lines									X

Fig. 5.26. Matrix of the correlations ($\hat{Q}$) among indicators in Fig. 5.9 (see Fig. 5.24). Correlations are identified by the symbol X. Already analysed indicators (that is indicators already included in the *Critical Few* set, or indicators correlated to them) are highlighted in dark grey.

This revised procedure is now applied to the *Help Desk* example (Sect. 5.3.1, Fig. 5.9). Table 5.15 shows the correlation matrix **Q**. The matrix ($\hat{Q}$) in Fig. 5.27 is obtained setting $k = 0.10$.

In the first step, we select all those indicators which are not correlated each other. In this specific case, each indicator has more than one correlation (see Fig. 5.27).

Then, the indicator with the maximum number of correlations is selected; it is the indicator "Uniformity of replies from different operators" (see Fig. 5.28).

	(I_1) Routing effectiveness	(I_2) Reply accuracy	(I_3) Uniformity of replies (different operators)	(I_4) Time for requests implementation	(I_5) Competence perception	(I_6) Answered calls percentage	(I_7) Courtesy of responses	(I_8) Security of data keeping	(I_9) Number of active lines
(I_1) Routing effectiveness	X	X	X	X	X			X	
(I_2) Reply accuracy	X	X	X		X				
(I_3) Uniformity of replies (from different operators)	X	X	X	X	X		X	X	
(I_4) Time for requests implementation	X		X	X	X				
(I_5) Competence perception	X	X	X	X	X				
(I_6) Answered calls percentage						X			X
(I_7) Courtesy of responses			X				X		
(I_8) Security of data keeping	X		X					X	
(I_9) Number of active lines						X			X

Fig. 5.27. Matrix of the correlations ($\hat{Q}$) among indicators in Fig. 5.9 (see Fig. 5.24). Threshold value (k) is 0.10 (see Table 5.15). Correlations are identified by the symbol X

	(I_1) Routing effectiveness	(I_2) Reply accuracy	(I_3) Uniformity of replies (different operators)	(I_4) Time for requests implementation	(I_5) Competence perception	(I_6) Answered calls percentage	(I_7) Courtesy of responses	(I_8) Security of data keeping	(I_9) Number of active lines
(I_1) Routing effectiveness	X	X	X	X	X			X	
(I_2) Reply accuracy	X	X	X		X				
(I_3) Uniformity of replies (from different operators)	X	X	X	X	X		X	X	
(I_4) Time for requests implementation	X		X	X	X				
(I_5) Competence perception	X	X	X	X	X				
(I_6) Answered calls percentage						X			X
(I_7) Courtesy of responses			X				X		
(I_8) Security of data keeping	X		X					X	
(I_9) Number of active lines						X			X

Fig. 5.28. Matrix of the correlations ($\hat{Q}$) among indicators in Fig. 5.9 (see Fig. 5.27). Threshold value (k) is 0.10 (see Table 5.15). Correlations are identified by the symbol X. Second step consists in identifying – among the remaining indicators – those with the maximum number of correlations (in light grey)

	(I₁) Routing effectiveness	(I₂) Reply accuracy	(I₃) Uniformity of replies (different operators)	(I₄) Time for requests implementation	(I₅) Competence perception	(I₆) Answered calls percentage	(I₇) Courtesy of responses	(I₈) Security of data keeping	(I₉) Number of active lines
(I_1) Routing effectiveness	X	X	X	X	X			X	
(I_2) Reply accuracy	X	X	X		X				
(I_3) Uniformity of replies (from different operators)	X	X	X	X	X		X	X	
(I_4) Time for requests implementation	X		X	X	X				
(I_5) Competence perception	X	X	X	X	X				
(I_6) Answered calls percentage						X			X
(I_7) Courtesy of responses			X				X		
(I_8) Security of data keeping	X		X					X	
(I_9) Number of active lines						X			X

Fig. 5.29. Matrix of the correlations ($\hat{Q}$). Threshold value (k) is 0.10. The third step consists in identifying – among the remaining indicators (in light grey) – the indicator with the maximum number of correlations. Dark grey columns identify indicators already included in the *Critical Few* set, or correlated to them

	(I₁) Routing effectiveness	(I₂) Reply accuracy	(I₃) Uniformity of replies (different operators)	(I₄) Time for requests implementation	(I₅) Competence perception	(I₆) Answered calls percentage	(I₇) Courtesy of responses	(I₈) Security of data keeping	(I₉) Number of active lines
(I_1) Routing effectiveness	X	X	X	X	X			X	
(I_2) Reply accuracy	X	X	X		X				
(I_3) Uniformity of replies (from different operators)	X	X	X	X	X		X	X	
(I_4) Time for requests implementation	X		X	X	X				
(I_5) Competence perception	X	X	X	X	X				
(I_6) Answered calls percentage						X			X
(I_7) Courtesy of responses			X				X		
(I_8) Security of data keeping	X		X					X	
(I_9) Number of active lines						X			X

Fig. 5.30. Matrix of the correlations ($\hat{Q}$) among indicators in Fig. 5.9. Threshold value (k) is 0.10. Fourth step consists in identifying – among the remaining indicators – the indicator with the maximum number of correlations. Since there are no other correlated indicators, the procedure stops

By removing the selected indicator and the ones to which it is correlated, we obtain the matrix shown in Fig. 5.29. Among the remaining indicators, those with the maximum number of correlations are: "Answered calls percentage" and "Number of active lines". The first one is selected.

Removing the selected indicator and those ones correlated to it, we obtain the matrix shown in Fig. 5.30. Since there are no other correlated indicators, the procedure stops.

The obtained Critical Few set includes indicators I_3 and I_6: "Uniformity of replies (from different operators)" and "Answered calls percentage".

As shown in Fig. 5.9, this set does not guarantee a complete covering. As a consequence, threshold k is increased to 0.20, obtaining the matrix ($\hat{Q}$) shown in Fig. 5.31. Even with this new threshold value, each indicator has more than one correlation. The first indicator included in the *Critical Few* set is "Routing effectiveness", since it has the maximum number of correlations (see Fig. 5.32).

	(I_1) Routing effectiveness	(I_2) Reply accuracy	(I_3) Uniformity of replies (different operators)	(I_4) Time for requests implementation	(I_5) Competence perception	(I_6) Answered calls percentage	(I_7) Courtesy of responses	(I_8) Security of data keeping	(I_9) Number of active lines
(I_1) Routing effectiveness	X	X	X	X	X			X	
(I_2) Reply accuracy	X	X	X		X				
(I_3) Uniformity of replies (from different operators)	X	X	X		X		X		
(I_4) Time for requests implementation	X			X					
(I_5) Competence perception	X	X	X		X				
(I_6) Answered calls percentage						X			X
(I_7) Courtesy of responses				X			X		
(I_8) Security of data keeping	X							X	
(I_9) Number of active lines						X			X

Fig. 5.31. Matrix of the correlations ($\hat{Q}$) among indicators in Fig. 5.9. Threshold value (k) is 0.20 (see Table 5.15). Correlations are identified by the symbol X

Matrix in Fig. 5.33 is obtained removing the selected indicator and those ones correlated to it.

	(I₁) Routing effectiveness	(I₂) Reply accuracy	(I₃) Uniformity of replies (different operators)	(I₄) Time for requests implementation	(I₅) Competence perception	(I₆) Answered calls percentage	(I₇) Courtesy of responses	(I₈) Security of data keeping	(I₉) Number of active lines
(I₁) Routing effectiveness	X	X	X	X	X			X	
(I₂) Reply accuracy	X	X	X		X				
(I₃) Uniformity of replies (from different operators)	X	X	X		X		X		
(I₄) Time for requests implementation	X			X					
(I₅) Competence perception	X	X	X		X				
(I₆) Answered calls percentage						X			X
(I₇) Courtesy of responses			X				X		
(I₈) Security of data keeping	X							X	
(I₉) Number of active lines						X			X

Fig. 5.32. Matrix of the correlations ($\hat{Q}$). Threshold value (k) is 0.20. Second step consists in identifying – among the remaining indicators (in light grey) – the indicator with the maximum number of correlations

	(I₁) Routing effectiveness	(I₂) Reply accuracy	(I₃) Uniformity of replies (different operators)	(I₄) Time for requests implementation	(I₅) Competence perception	(I₆) Answered calls percentage	(I₇) Courtesy of responses	(I₈) Security of data keeping	(I₉) Number of active lines
(I₁) Routing effectiveness	X	X	X	X	X			X	
(I₂) Reply accuracy	X	X	X		X				
(I₃) Uniformity of replies (from different operators)	X	X	X		X		X		
(I₄) Time for requests implementation	X			X					
(I₅) Competence perception	X	X	X		X				
(I₆) Answered calls percentage						X			X
(I₇) Courtesy of responses			X				X		
(I₈) Security of data keeping	X							X	
(I₉) Number of active lines						X			X

Fig. 5.33. Matrix of the correlations ($\hat{Q}$). Threshold value (k) is 0.20. The third step consists in identifying – among the remaining indicators (in light grey) – the indicator with the maximum number of correlations. The dark grey columns identify the indicators which have already been analysed (indicators included in the *Critical Few* set, or indicators correlated to them)

Among the remaining indicators, those with the maximum correlations number are: I_6 ("Answered calls percentage") and I_9 ("Number of active lines") (see Fig. 5.33). I_6 is selected. The remaining indicator – I_7 ("Courtesy of responses") – is included in the *Critical Few* set.

The new *Critical Few* set is composed by indicators I_1, I_6 and I_7. Again, as shown in Fig. 5.9, the new set does not guarantee a complete covering. It is interesting to notice that in choosing indicator I_9, instead of I_6, the resulting set does not guarantee a complete covering, either.

Threshold is increased to the value $k = 0.75$, in order to identify an indicators set which guarantees a complete covering. New matrix $\hat{\mathbf{Q}}$ is shown in Fig. 5.34. In this specific case, some indicators are not correlated with other ones: I_4 ("Time for requests implementation"), I_6 ("Answered calls percentage"), I_7 ("Courtesy of responses"), I_8 ("Security of data keeping") and I_9 ("Number of active lines"). These indicators are removed from the matrix and included in the *Critical Few* set.

	(I_1) Routing effectiveness	(I_2) Reply accuracy	(I_3) Uniformity of replies (different operators)	(I_4) Time for requests implementation	(I_5) Competence perception	(I_6) Answered calls percentage	(I_7) Courtesy of responses	(I_8) Security of data keeping	(I_9) Number of active lines
(I_1) Routing effectiveness	X	X	X		X				
(I_2) Reply accuracy	X	X	X		X				
(I_3) Uniformity of replies (from different operators)	X	X	X						
(I_4) Time for requests implementation				X					
(I_5) Competence perception	X	X			X				
(I_6) Answered calls percentage						X			
(I_7) Courtesy of responses							X		
(I_8) Security of data keeping								X	
(I_9) Number of active lines									X

Fig. 5.34. Matrix of the correlations ($\hat{\mathbf{Q}}$) among indicators in Fig. 5.9. Threshold value (*k*) is 0.75 (see Fig. 5.24 and Table 5.15). Correlations are identified by the symbol X. The first step consists in identifying the indicator with no correlations (in light grey)

The second step consists in identifying the indicator with the maximum number of correlations. In this case, two indicators are found: I_1 ("Routing effectiveness") and I_2 ("Reply accuracy") (see Fig. 5.35). The first one is included in the *Critical Few* set.

	(I₁) Routing effectiveness	(I₂) Reply accuracy	(I₃) Uniformity of replies (different operators)	(I₄) Time for requests implementation	(I₅) Competence perception	(I₆) Answered calls percentage	(I₇) Courtesy of responses	(I₈) Security of data keeping	(I₉) Number of active lines
(I₁) Routing effectiveness	X	X	X		X				
(I₂) Reply accuracy	X	X	X		X				
(I₃) Uniformity of replies (from different operators)	X	X	X						
(I₄) Time for requests implementation				X					
(I₅) Competence perception	X	X			X				
(I₆) Answered calls percentage						X			
(I₇) Courtesy of responses							X		
(I₈) Security of data keeping								X	
(I₉) Number of active lines									X

Fig. 5.35. Matrix of the correlations ($\hat{Q}$). Threshold value (k) is 0.75. The second step consists in identifying – among the remaining indicators (in light grey) – the indicator with the maximum number of correlations. Dark grey columns identify indicators already included in the *Critical Few* set, or indicators correlated to them

	(I₁) Routing effectiveness	(I₂) Reply accuracy	(I₃) Uniformity of replies (different operators)	(I₄) Time for requests implementation	(I₅) Competence perception	(I₆) Answered calls percentage	(I₇) Courtesy of responses	(I₈) Security of data keeping	(I₉) Number of active lines
(I₁) Routing effectiveness	X	X	X		X				
(I₂) Reply accuracy	X	X	X		X				
(I₃) Uniformity of replies (from different operators)	X	X	X						
(I₄) Time for requests implementation				X					
(I₅) Competence perception	X	X			X				
(I₆) Answered calls percentage						X			
(I₇) Courtesy of responses							X		
(I₈) Security of data keeping								X	
(I₉) Number of active lines									X

Fig. 5.36. Matrix of the correlations ($\hat{Q}$). Threshold value (k) is 0.75 (see Fig. 5.24). The third step consists in identifying – among the remaining indicators (in light grey) – the indicator with the maximum number of correlations. Since there are not other correlated indicators, the procedure stops

Removing the new selected indicator and those with which it has a correlation, we obtain the matrix shown in Fig. 5.36. Since there are no other correlated indicators, the procedure stops.

The *Critical Few* set is composed by indicators I_1, I_4, I_6, I_7, I_8 and I_9. In this case, the set guarantees a complete covering and the procedure stops.

The illustrated procedure guarantees a complete "covering" of the representation-targets, but it is not able to minimize the selected indicators. In the worst case, iterations stop when threshold k value is 1, and all the indicators are included in the *Critical Few* set.

A second drawback is given by the numerical encoding of the Relationship Matrix coefficients, since they are defined only in qualitative terms (order relation only). It is important to notice that the encoding is based on arbitrary values (see Fig. 5.22 and 5.23) and it may generate significant distortions in the *Critical Few* selection.

In conclusion, these methods represent only a part of the heuristic methodologies which can be used to select *Critical Few* indicators. The level of refinement of the model typically depends on the quality of the available information.

5.6   Implementing a system of performance indicators

To develop a performance measurement system, a proper support organization is needed. This section deals with this aspect, introducing some guidelines which can be helpful for the system implementation. Description is supported by some practical examples.

The following operational steps contribute to the organization improvement for the development of a performance measurement system.

Step 1: *Establishing the working group which will activate the performance measurement system.*

Step 2: *Defining a proper «terminology» within the organization.*
In general, in addition to the classifications presented in Sect. 1.4 and 4.4, indicators can be divided into five categories:

- *Input Indicators*: used to understand the human and capital resources used to produce the outputs and outcomes.
- *Process Indicators*: used to understand the intermediate steps in producing a product or service. In the area of training for example, a process indicator could be the number of training courses completed as scheduled.

- *Output Indicators*: used to measure the product or service provided by the system or organization and delivered to customers/users. An example of a training output would the number of people trained.
- *Outcome Indicators*: evaluate the expected, desired, or actual result(s) to which the outputs of the activities of a service or organization have an intended effect. For example, the outcome of safety training might be improved safety performance as reflected in a reduced number of injuries and illnesses in the workforce.
- *Impact Indicators*: measure the direct or indirect effects or consequences resulting from achieving program goals. An example of an impact is the comparison of actual program outcomes with estimates of the outcomes that would have occurred in the absence of the program

A second possible classification is based on the temporal moment in which measurements are performed. These types of measurements are defined below:

- *Lagging Measurements*: measure performance after the fact. Project cost performance is an example of a lagging indicator used to measure program performance.
- *Leading Measurements*: are more predictive of future performance. They include, for example, measurements such as procedural violations, or estimated cost based on highly correlated factors.
- *Behavioural Measurements*: measure the underlying culture or attitude of the personnel or organization being measured. A classical example is given by employee satisfaction questionnaires.

Step 3: *Design general criteria.*
Here are some general criteria to consider when developing a performance measurement system:

- keep the number of performance indicators to a minimum. For any program, there are a large number of potential performance indicators. It is important to identify a limited number of "critical indicators";
- process objectives must be understandable and must be developed clearly. Experience has shown that performance measurement systems frequently fail because the respective parties do not have a common understanding regarding the purpose and concepts of the performance measurement system;
- determine if the cost of the performance indicator is worth the gain. Sometimes the cost of obtaining a measurement may outweigh any added value resulting from the measurement (see Sect. 4.6);

- assure that the measure is comprehensive. In developing performance indicators, consider measuring positive performance as well as minimizing possible negative side-effects of the program. (see Sect. 4.6);
- consider performing a risk evaluation. Organizations should consider performing a risk evaluation of the organization to determine which specific processes are most critical to organizational success or which processes pose the greatest risk to successful mission accomplishments;
- place greater emphasis on measuring the risk produced by the use of a particular performance indicator, both for short and long-term;
- consider the weight of conflicting performance measures. For example, an objective of high productivity may conflict with an objective for a high quality product (property of *non counter-productivity*, Sect. 4.6.1);
- develop consistent performance measures that promote teamwork. Performance measures should be designed to maximize teamwork between different organizational elements. The performance measures for different levels of an organization should be generally consistent with each other, from top to bottom and across the hierarchy. The risks of suboptimization should be determined when setting performance indicators. An example of suboptimization is: a technical group, in its haste to complete a project, prepares an incomplete, error-filled specification which prevents the contractor from completing the project on time and it results in increased costs (property of *non counter-productivity*, Sect. 4.6.1).

Step 4: *How to check performance measures.*

After having developed a system of performance indicators, it is important to check/test it. Here are several possible checks/tests.

SMART test (University of California 1998)

- **S** (*Specific*): is the measure clear and focused, so it avoids misinterpretation? It should include measurement assumptions and definitions, and should be easily interpreted.
- **M** (*Measurable*): can the measure be quantified and compared to other data? It should allow for meaningful statistical analysis.
- **A** (*Attainable*): is the measure achievable, reasonable, and credible under expected conditions?
- **R** (*Realistic*): does the measure fit into the organization's constraints? Is it cost-effective?
- **T** (*Timely*): is the measurement doable within the given time frame?

The *"Three Criteria"* test (Performance-Based Management Special Interest Group 2001)
Another test to which performance indicators should be subjected includes the satisfaction of three broad criteria:

- *Strategic Criteria* – do the measures enable strategic planning and then drive the deployment of the actions required to achieve objectives and strategies? Do the measures align behaviour and initiatives with strategy, and focus the organization on its priorities?
- *Quantitative Criteria* – do the measures provide a clear understanding of progress toward objectives and strategy as well as the current status, rate of improvement, and probability of achievement? Do the measures identify gaps between current status and performance aspirations, thereby highlighting improvement opportunities?
- *Qualitative Criteria* – are the measures perceived as valuable by the organization and the people involved with the indicators?

The Treasury Department Criteria test (U.S. Department of the Treasury 1994)
The test is based on the following general verification criteria:

1. Data criteria - data availability and reliability can impact the selection and development of performance measures.

- *Availability*: are the data currently available? If not, can the data be collected? Are better indicators available using existing data? Are there better indicators that we should be working towards, for which the data are not currently available?
- *Accuracy*: are the data sufficiently reliable? Are there biases, exaggerations, omissions, or errors that are likely to make an indicator or measure inaccurate or misleading? Are the data verifiable and auditable?
- *Timeliness*: are the data timely enough for evaluating program performance? How frequently are the data collected and/or reported (e.g., monthly vs. annually)?
- *Security*: are there privacy or confidentiality concerns that would prevent the use of these data by concerned parties?
- *Costs of data collection*: are there sufficient resources (e.g., expertise, computer capability or funds) available for data collection? Is the collection of the data cost-effective?

2. Indicator criteria

- *Validity*: does the indicator address financial or program results? Can changes in the value of the indicator be clearly interpreted as desirable or undesirable? Does the indicator clearly reflect changes in the program? Is there a sound, logical relationship between the program and what is being measured, or are there significant uncontrollable factors?

- *Uniqueness*: does the information conveyed by one indicator duplicate information provided by another (redundancies or correlations)?
- *Evaluation*: are there reliable benchmark data, standards, or alternative frames of reference for interpreting the selected performance indicators?

3. Measurement system criteria

- *Balance*: is there a balance between input, output, and outcome indicators, and productivity or cost-effectiveness indicators? Does the mix of indicators offset any significant bias in any single indicator?
- *Completeness*: are all major programs and major components of programs covered? Does the final set of indicators cover the major goals and objectives (concept of exhaustiveness – Sect. 4.6.2)
- *Usefulness*: will management use the system to effect change based on the analysis of the data? Are there incentives for management to use the data after they are collected? Does management have the resources to analyze the results of the system? Is management trained to use and interpret the data? Are management reports "user-friendly" - that is, clear and concise?

It is interesting to notice how some of these criteria overlap with the properties discussed in Chap. 4, although they do not provide an organic procedure to test them.

Step 5: *Benchmarking with other organizations' performance measuring systems.*

The point here is to eliminate "reinvent the wheel" and, thus, save time and resources not repeating errors made by other organizations.

5.6.1 Examples of developing performance measures

In literature, many different approaches for developing performance measurement systems have been proposed during the years. The following sections present two emblematic approaches, to give an idea of the actions needed to develop measures of performance:

- The Auditor General of Canada approach;
- The DOE/NV approach (U.S. Department of Energy/Nevada Operations 1994).

The Auditor General of Canada approach

This approach comes from the document "Developing Performance Measures for Sustainable Development Strategies", produced by the Office of the Auditor General of Canada and the Commissioner of the Environment and Sustainable Development (http://www.oag-bvg.gc.ca). It is designed to assist work units within departments in developing objectives and measures that contribute to achieving the department's strategic objectives for sustainable development.

The idea that there must be direct linkages between the strategic objectives set by the department and the objectives, action plans and measures of each of its work unit forms the basis of this approach.

The methodology is organized into two main parts:

1. Definition of program-level objectives that contribute to strategic objectives. The activity contains five work steps (1 through 5).
2. Establish performance measures. Description is divided into four work steps (6 through 9). These steps are intended to assist users in establishing sound performance measures which correspond to their objectives as well as accountability and resource requirements for implementation.

PART 1

A performance framework brings structure to performance planning and clarifies the connection between activities, outputs and results. A good performance framework will address the following questions relative to the objectives specified in the department's strategic plan:

- *WHY* is your program relevant to the strategic objective?
  This question relates to the long-term, sustainable development result(s) that the program can reasonably be expected to produce in support of a strategic objective.
- *WHO* do you want to reach?
  (target groups, stakeholders)
- *WHAT* results do you expect to achieve?
  This question relates to the short-term (or intermediate) result(s) of program activities or outputs that are believed to contribute to achieving the long-term results.
- *HOW* are you going to achieve your objectives?
  This question relates to program inputs, processes, activities and outputs.

Step 1: *Confirm program role*
Defining the role that the program is intended to fulfil with respect to strategic objectives provides a basis for establishing program targets and

performance indicators. Table 5.16 – linking program activities to strategic objectives – has been developed to support this task. As an example, Table 5.16 refers to a program for the reduction of daily defectiveness in an automotive exhaust-systems production plant (see Sect. 3.3).

Table 5.16. Linking program activities to strategic objectives. The example data refer to a program for the daily defectiveness reduction in an automotive exhaust-systems production plant (see Sect. 3.3)

Main activities or outputs of program	Contributes to/detracts from a specify the strategic objectives	Strategic objectives or outcomes to which the program activity or output contributes
Activity 1: Personnel training	It increases personnel's competence, skill and participation	Reduction of human errors
Activity 2: Programmed preventive maintenance	It increases factory system reliability	Reduction of the system's causes of malfunction/break-down
Activity 3: Introduction of a control system	It facilitates the quick detection of the production process faults	Process control

Step 2: *Identify the key program activities and outputs*

This step is essential to ensure that program managers and staff focus on key issues that contribute to the achievement of organizational strategy. Table 5.17 schematizes the procedure. Relationships among activities/outputs and objectives are encoded using a three-level qualitative scale: High (H), Medium (M) or Low (L).

Relationships in Table 5.17 are aggregated and synthesized with the aim of evaluating the overall impact of the activities/outputs onto objectives (see last column). This operation can be carried out by the use of different aggregation techniques (Franceschini et al. 2005).

In this case, to simplify the analysis, we consider – for each activity/output – the minimum relationships level. L_{AGG}, the aggregated importance, is given by the minimum value of the relationships with the objectives:

$$L_{AGG} = \min_{o_i \in O}\{L(o_i)\} \qquad (5.8)$$

being O the whole set of objectives.

Step 3: *Identify program stakeholders and issues*

In order to formulate a set of strategic objectives, it is essential to identify: who the program activities and outputs are intended to serve, influence or target, who the other principal groups affected are and how they are affected. For significant program activities/outputs, Table 5.18 identifies their link to stakeholders and issues.

Table 5.17. Identifying the key program activities and outputs. The relationship is encoded using a three-level qualitative scale: High (H), Medium (M) or Low (L). Last column includes the aggregated importance related to each activity/output. Activities/outputs and strategic objectives are defined in Table 5.16

Program activities and outputs	Strategic objective			Aggregated importance $L_{AGG} = \min_{o_i \in O}\{L(o_i)\}$
	Reduction of human errors	Reduction of the system's malfunction/break-down	Process control	
Activity 1: personnel training	H	H	H	H
Output 1: personnel's competence increase	H	M	M	M
Activity 2: programmed preventive maintenance	L	H	L	L
Output 2: reduction of the number of system malfunctions/break-downs	L	H	M	L
Activity 3: introduction of a control system	H	H	H	H
Output 3: reduction of the time to detect the production process faults	M	M	H	M

Step 4: *Identify what the program aims to accomplish*

Table 5.19 establishes a connection between activities/outputs – in order of significance – and medium/long-term strategic objectives.

Step 5: *Identify responses and performance requirements*

Performance objectives must be defined in operational terms in order to be managed effectively. Table 5.20 establishes a connection between performance requirements and desired results (objectives' operationalization).

PART 2

The next four steps are intended to assist the user in establishing sound performance measures as well as accountability and resource requirements for implementation.

Table 5.18. Identifying key issues and affected stakeholder groups with reference to the activities/outputs defined in Table 5.17. Activities and outputs are listed in order of significance (High, Medium, Low)

Main program activities and outputs in order of significance (H, M, L)	Key issues		Stakeholder groups (affected parties)	
	Desired program effects	Undesired program effects	Positively affected	Negatively affected
Activity 1 (H): personnel training	reduction of defectiveness due to human errors; personnel participation	needed resources for personnel training part of the working hours are spent for training	process operators	production director
Activity 3 (H): Introduction of a control system	detection of the causes of faults	increase of the process time personnel required for the conduction of the process control	process operators, process leaders	production director
Output 1 (M): personnel's competence increase	increase of the personnel's competence and skill	personnel's expectation of career advancements	process operators	none
Output 3 (M): reduction of the time to detect the production process faults	quick solution of the problems	none	process operators, process leaders	none
Activity 2 (L): programmed preventive maintenance	increase of the process reliability	resources needed for the analysis of the process reliability resources needed for maintenance work production should be periodically stopped	process leaders	none
Output 2 (L): reduction of the number of system malfunctions/break-downs	increase of the process reliability	none	process leaders	none

Step 6: *Identify potential performance measures*

Performance measurement is required to understand the gap between actual and expected levels of achievement and when corrective action may be warranted. The results indicated by a performance measure will be generally compared to the expectations specified by a performance target.

Table 5.19. Defining results with reference to the activities/outputs defined in Table. 5.18. Activities and outputs are listed in order of significance (High, Medium, Low)

Main program activities and outputs in order of significance	Desired results (objectives)	
	Medium-term intermediate	Long-term strategic
Activity 1 (H): personnel training	Personnel's awareness of the process issues	Highly qualified personnel for each process activity
Activity 3 (H): Introduction of a control system	Establishment of a process control system	Process completely under control
Output 1 (M): personnel's competence increase	50% (at least) of the operators are qualified	All the operators are qualified
Output 3 (M): reduction of the time to detect the production process faults	Establishment of a process faults detection system	Enhancement (in terms of quickness) of the process faults detection system
Activity 2 (L): programmed preventive maintenance	Analysis of the factory system reliability	Efficient system for programmed maintenance
Output 2 (L): reduction of the number of system malfunctions/break-downs	Significant reduction of system malfunctions/break-downs	Total absence of system malfunctions/break-downs

Table 5.20. Defining performance requirements with reference to the desired results (objectives). The analysis concerns the long-term strategic results only, but it can be extended to the medium-term

Objective(s)	New or modified activities, outputs or other program response(s) necessary to achieve the objective(s)	Performance requirements relative to each activity, output or other response necessary to achieve the desired results
Highly qualified personnel for each process activity	*Activity 1* (H): personnel training	Organization of a training plan (selection of the matters, teachers' recruitment, etc...)
Process completely under control	*Activity 3* (H): Introduction of a control system	Establishment of the control procedures, tools (e.g. control charts) and monitored variables
All the operators are qualified	*Output 1* (M): personnel's competence increase	Establishment of a system to evaluate personnel's competence
Enhancement (in terms of quickness) of the process faults detection system	*Output 3* (M): reduction of the time to detect the production process faults	Establishment of a system for quick data collecting and analysis
Efficient system for programmed maintenance	*Activity 2* (L): programmed preventive maintenance	Establishment of a programmed maintenance plan
Total absence of system malfunctions/break-downs	Output 2 (L): reduction of the number of system malfunctions/break-downs	Establishment of a system malfunctions/break-downs monitoring system (analysis of reliability)

Table 5.21. Identifying potential performance measures, with reference to the performance requirements defined in Table 5.20

Objective(s)	Activities, outputs or other program responses	Performance requirements	Potential performance measure(s)
Highly qualified personnel for each process activity	*Activity 1* (H): personnel training	Organization of a training plan (selection of the matters, teachers' recruitment, etc...)	Percentage of highly qualified employees for each process activity
Process completely under control	*Activity 3* (H): Introduction of a control system	Establishment of the control procedures, tools (e.g. control charts) and monitored variables	Number of defects identified during a specific time period, due to a weak process control system
All the operators are qualified	*Output 1* (M): personnel's competence increase	Establishment of a system to evaluate personnel's competence	Percentage of employees considered "very competent", according to the evaluation system
Enhancement (in terms of quickness) of the process faults detection system	*Output 3* (M): reduction of the time to detect the production process faults	Establishment of a system for quick data collecting and analysis	Time between the fault occurrence and its detection
Efficient system for programmed maintenance	*Activity 2* (L): programmed preventive maintenance	Establishment of a programmed maintenance plan	Number of system malfunctions/breakdowns in a specific time period, due to a lack of programmed maintenance
Total absence of system malfunctions/break-downs	*Output 2* (L): reduction of the number of system malfunctions/break-downs	Establishment of a system malfunctions/break-downs monitoring system (analysis of reliability)	Total number of system malfunctions/break-downs in a specific time period

Performance measures are an important source of feedback for effective management. Table 5.21 establishes a connection among desired results, performance requirements, and potential performance measures.

Step 7: *Establish information capabilities and a baseline for each measure*

Understanding what information is currently available to the organization as well as the organization's capabilities for collecting and analyzing information is an important first step in the selection of performance measures. Table 5.22 can be useful in order to reach this purpose.

Step 8: *Assess the adequacy of performance measurements*
Once a list of candidate performance measurements has been developed, the next step is to select a set of performance measurements that are suitable for tracking performance toward specified objectives.

Table 5.22. Data collecting scheme. Performance measurements (indicators) are those defined in Table 5.21

Potential performance measure(s)	Units	Initial value
Percentage of highly qualified employees for each process activity	%	50
Number of defects identified during a specific time period, due to a weak process control system	number / month	0
Percentage of employees considered as competent, according to the evaluation system	%	50
Time between the fault occurrence and its detection	hours	24
Number of system malfunctions/break-downs during a specific time period, due to a lack of programmed maintenance	number / month	0
Total number of system malfunctions/break-downs in a specific time period	number / month	5

Below is a selection of criteria for performance measurements:

- *Meaningful*
  - understandable (clearly and consistently defined, well explained, measurable, with no ambiguity);
  - relevant (relates to objectives, significant and useful to the users, attributable to activities);
  - comparable (allows comparison over time or with other organizations, activities or standards).
- *Reliable*
  - accurately represents what is being measured (valid, free from bias, etc...);
  - data required can be replicated (verifiable);
  - data and analysis are free from error;
  - not susceptible to manipulation;
  - balances (complements) other measurements.
- *Practical*
  - feasible financially;
  - feasible to get timely data.

It is interesting to compare these requirements with those presented in the taxonomy in Chap. 4.

Table 5.23 helps to select the most suitable indicators.

Table 5.23. Scheme for selecting the most suitable indicators. Performance measurements (indicators) are those defined in Table 5.21

Performance Measurements	Meaningful			Reliable	Practical
	Under-standable	Rele-vant	Com-parable		
Number of defects identified during a specific time period, due to a weak process control system	N	N	N	N	Y
Time between the fault occurrence and its detection	Y	Y	Y	N	Y
Percentage of highly qualified employees for each process activity	N	Y	N	Y	Y
Percentage of employees considered "very competent", according to the evaluation system	Y	N	N	Y	Y
Number of system malfunctions/break-downs during a specific time period, due to a lack of programmed maintenance	N	Y	N	N	N
Total number of system malfunctions/break-downs in a specific time period	Y	Y	Y	Y	Y

Step 9: *Establish accountability and resources for implementation*

An accountability system formalizes the relationship between results, outputs, activities, and resources. It allows people to see how their work contributes to the success of the organization and clarifies expectations for performance. Table 5.24 and Table 5.25 provide a reference scheme which may be helpful.

According to the *"Process Auditor"* method, a set of performance measures should support a broader explanation of performance results for managers and executives and for internal and external stakeholders. Performance information explains how the resources committed to specific initiatives for achieving performance objectives do or do not allow the achievement of specified results. The method has a strong "constitutive" connotation.

Table 5.24. Establishing accountability for the performance measurement system. Desired results and activities/outputs have been defined in Table 5.20. Performance measures (indicators) are taken from Table 5.22

Desired results (objectives)	Responsible party(s) for achieving objective	Activities, outputs or other responses necessary to meet objectives	Responsible party(s) for managing activities or outputs and meeting the requirements	Performance measurement(s)	Responsible party(s) for evaluating measurements
Highly qualified personnel for each process activity	Personnel manager	*Activity 1* (H): personnel training	Personnel manager (training area)	Percentage of highly qualified employees for each process activity	Statistics operators (personnel area)
Process completely under control	Production manager	*Activity 3* (H): introduction of a control system	Quality manager	Number of defects identified during a specific time period, due to a weak process control system	Process control operators
All the operators are qualified	Personnel manager	*Output 1* (M): personnel's competence increase	Personnel manager (training area)	Percentage of employees considered "very competent", according to the evaluation system	Statistics operators (personnel area)
Enhancement (in terms of quickness) of the process faults detection system	Production manager	*Output 3* (M): reduction of the time to detect the production process faults	Quality manager	Time between the fault occurrence and its detection	Process control operators
Efficient system for programmed maintenance	Production manager	*Activity 2* (L): programmed preventive maintenance	Manager of the maintenance	Number of system malfunctions/breakdowns during a specific time period, due to a lack of programmed maintenance	Maintenance operators
Total absence of system malfunctions/breakdowns	Production manager	*Output 2* (L): reduction of the number of system malfunctions/breakdowns	Quality manager	Total number of system malfunctions/break-downs in a specific time period	Maintenance operators

Table 5.25. Identifying resource requirements for implementation. Data are taken from Table 5.20.

Program objectives	Activities/ outputs or other responses necessary to meet objectives	Resource requirements		
		Human	Financial	Other
Highly qualified personnel for each process activity	*Activity 1* (H): personnel training	2	50·000 € per year	
Process completely under control	*Activity 3* (H): Introduction of a control system	2	65·000 € per year	
All the operators are qualified	*Output 1* (M): personnel's competence increase			
Enhancement (in terms of quickness) of the process faults detection system	*Output 3* (M): reduction of the time to detect the production process faults			
Efficient system for programmed maintenance	*Activity 2* (L): programmed preventive maintenance	3	100·000 € per year	
Total absence of system malfunctions/break-downs	*Output 2* (L): reduction of the number of system malfunctions/break-downs			

The DOE/NV approach

This section presents a methodology for establishing a performance measurement system proposed by DOE/NV (U.S. Department of Energy 1994; Performance-Based Management Special Interest Group 1995). The methodology is applied to a practical case study.

The Communications & Information Management Company provides, communications and information management services. The company's warehouse, part of the Property Management Division, provides storage and excess services for company property in the custody of 25 divisions. The warehouse department has a staff of 10 personnel: a warehouse supervisor, four property specialists, one property clerk, three drivers, and one data entry clerk. The warehouse makes approximately 50 pickups per week at company locations that include remote areas.

To request services from the warehouse, a division customer telephones the warehouse property clerk requesting a pick-up of goods for storage or excess. The customer provides the clerk with the goods identification number or serial number for each good to be picked up and brought to the warehouse. There are typically one to twenty goods per pick-up. If a pick-up date is not requested by the customer, a date will be provided to the customer by the property clerk. The property clerk completes a property transfer form, which reflects the date of the call, customer's name, division, location, property identification number and date scheduled for pick-up.

A goal of the warehouse is not to exceed three days from the date of the call to the time of the pick-up, unless a special date has been requested by the customer. The warehouse receives approximately ten calls per week for pick-ups on special dates. On the scheduled pick-up day, the assigned driver takes the transfer form to the designated location. The driver is responsible for ensuring each good matches the property identification numbers or serial numbers listed on the transfer form. After the truck is loaded, the driver obtains the customer's signature on the transfer form. The driver also signs the form and provides the customer with a copy acknowledging the receipt.

The driver returns to the warehouse, where a property specialist annotates the date on the transfer form, unloads the truck, and provides the data entry clerk with the signed copies of the form. The data entry clerk enters the information from the transfer form into the automated accountable property system and the transfer forms are then filed. The data entered are intended to transfer accountability from the division customer to the warehouse. At the end of the month, division customers receive a computer-generated property list indicating the accountable property in their location for which they are responsible. The customer reviews this report for accuracy. If the customer records do not agree with this listing, the customer calls the warehouse supervisor who logs the complaint with the following information: date of the call, division name, property location, date of the property list, and discrepancies. The supervisor assigns a property specialist to resolve these discrepancies.

The group is responsible for many processes, such as delivering property, conducting inventory, etc. For purposes of simplicity, the following description summarizes the operational steps to develop performance measurements for the process *"goods pick-up and storage"*. The work team involves the entire staff.

Step 1: *Process identification*
The first step consists in identifying process inputs, outputs, activities, and resources (see Sect. 5.3.2). Fig. 5.37 provides a flow chart representation of the analysed process.

This activity includes the identification of the process objectives and outputs (see Fig. 5.38). Process objectives are:

- a current, accurate goods list for customers;
- timely pick up and removal of goods.

Outputs are:

- a list of goods for customers;
- removal and storage of company goods.

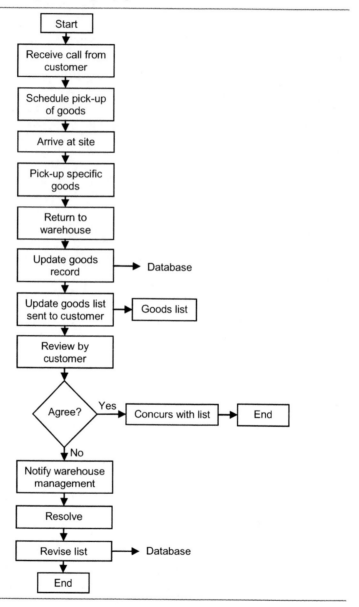

Fig. 5.37. Flow chart for the process of "goods pick-up and storage" (Performance-Based Management Special Interest Group 1995). With permission

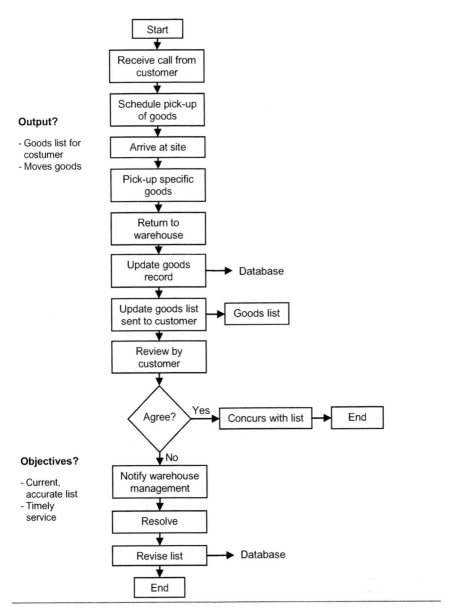

Fig. 5.38. Flow chart for the process of "goods pick-up and storage", supplied with the process objectives and outputs (Performance-Based Management Special Interest Group 1995). With permission

Step 2: *Identification of critical activity to be measured*

The next step is to determine how objectives will be met. In this case, work team identifies two sets of critical activities that needed to be watched closely and acted on, if performance is less than the desired goal. The reason why these were considered critical is because they are the sets of activities that produce process outputs. Two control points have been defined (see Fig. 5.39).

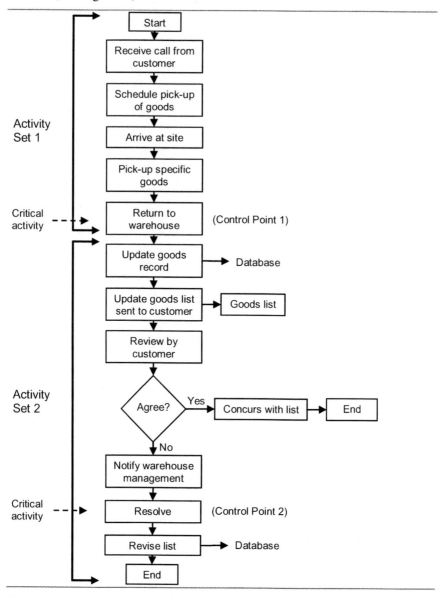

Fig. 5.39. Flow chart for the process of "goods pick-up and storage", supplied with the representation of the process critical activities and control points (Performance-Based Management Special Interest Group 1995). With permission

Step 3: *Establishing performance goals or standards*
For each control point selected for measurement, it is necessary to establish a performance goal or standard.

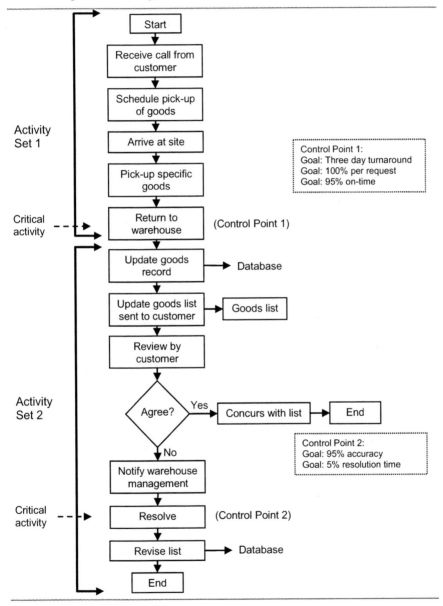

Fig. 5.40. Flow chart for the process of "goods pick-up and storage", supplied with performance goals related to critical activities (Performance-Based Management Special Interest Group 1995). With permission

Looking at critical activity 1 (*"Return to warehouse"*), three goals have been defined (see Fig. 5.40):

- three-day turnaround;
- scheduling pick-up per customer request;
- 95% on time pick-ups.

For critical activity 2 (*"Resolve discrepancies"*) (see Fig. 5.40):

- 98% goods list accuracy;
- no more than 5% of the time should be dedicated in resolving discrepancies.

Step 4: *Establish performance measurement(s)*
Performance measures represent the most important aspects of the process (see Sects. 4.6.1 and 4.6.2). For this reason, we need to identify specific performance measures for the two critical activities.

In particular, looking at **critical activity 1** (*"Return to warehouse"*), we define the following indicators (see Fig. 5.41):

- *Performance indicator 1-A*: Number of days elapsed from call to pick-up
    - Collected data: date of call for pick-up services, actual date of pick-up.
    - Means of collection: goods transfer form.
    - Frequency: weekly.

- *Performance indicator 1-B*: Percentage of specially scheduled pick-ups on time:

$$\frac{\text{number on-time special pick-ups}}{\text{number scheduled special pick-ups}} \times 100 \qquad (5.9)$$

    - Collected data: number of special pick-ups scheduled each week and number on time.
    - Means of collection: goods transfer form.
    - Frequency: weekly.

- *Performance indicator 1-C*: Percentage of on-time pick-ups (for all pick-ups):

$$\frac{\text{number on-time pick-ups}}{\text{total number of pick-ups}} \times 100 \qquad (5.10)$$

- Collected data: total number of pick-ups completed, total number on-time pick-ups.
- Means of collection: goods transfer form.
- Frequency: weekly.

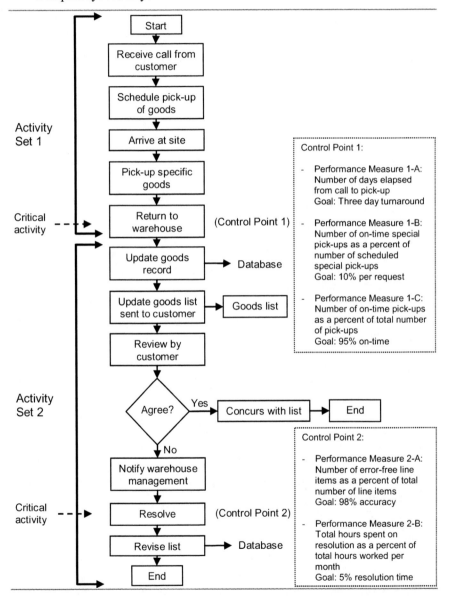

Fig. 5.41. Flow chart for the process of "goods pick-up and storage", supplied with performance indicators selected (Performance-Based Management Special Interest Group 1995). With permission

Considering the **critical activity 2** (*"Resolve discrepancies"*), we define the following indicators (see Fig. 5.41):

- *Performance indicator 2-A*:   Percentage of accuracy of monthly report:

$$\frac{\text{number of error-free line items}}{\text{total number of line items}} \times 100 \qquad (5.11)$$

  - Collected data: total number of line item entries generated each month on goods lists, number of errors detected (used to calculate number error-free).
  - Means of collection: goods list database, complaint log.
  - Frequency: monthly.

- *Performance indicator 2-B*: Percentage of time spent resolving goods list problems:

$$\frac{\text{total hours spent on resolutions}}{\text{total hours worked per month}} \times 100 \qquad (5.12)$$

  - Collected data: total number that the four goods specialists spend on problem resolution each month, total hours worked by the goods specialists each month.
  - Means of collection: special job number to add to the time card to track the time spent resolving goods list problems.
  - Frequency: monthly.

Step 5: *Identify responsible party(ies)*
The next step consists in identifying responsible parties for collecting the data, analyzing/reporting actual performance, comparing actual performance to goal/standard, determining if corrective actions are necessary, and making changes.

In this specific case, for the two critical activities, two goods specialists are responsible for collecting, interpreting, and providing feedback on the data. The warehouse supervisor is responsible for making decisions and taking action (see Fig. 5.42).

Step 6: *Collect data*
Data collection is much more than simply writing things down and then analyzing everything after a period of time. Even the best of measurement systems may fail because of poor data collection. Several preliminary analyses should be conducted in order to determine if the measurement system is functioning as designed, that the frequency of data collection is

appropriate, and to provide feedback to the data collectors with respect to any adjustments in the system.

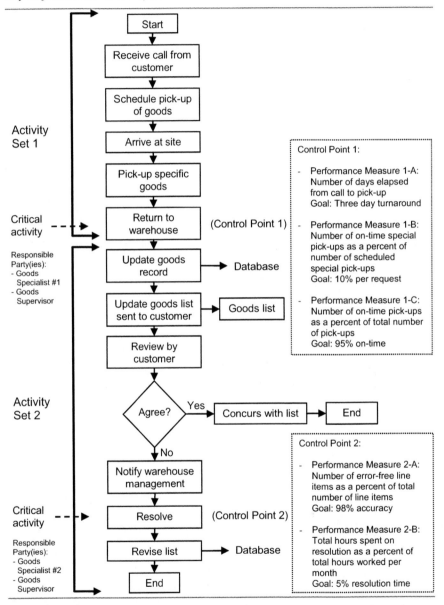

Fig. 5.42. Flow chart for the process of "goods pick-up and storage", supplied with the identification of the responsible parties for the two critical activities (Performance-Based Management Special Interest Group 1995). With permission

In this specific case two control points have been identified. The first control point covers flow process activity numbers 2, 4, and 5. The second control point covers activity numbers 11 and 12 (Fig. 5.43).

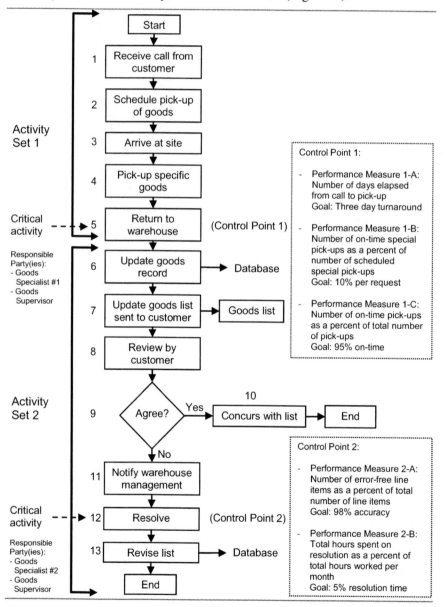

Fig. 5.43. Flow chart for the process of "goods pick-up and storage", supplied with activities indexing (Performance-Based Management Special Interest Group 1995). With permission

For the first control point, the use of an existing goods transfer form, already in use for recording the data, is supposed to be the most efficient means for collecting the necessary information:

- for activity 2: the date the customer places the request and the scheduled date for the pick-up;
- for activities 4 and 5: the date the property is actually picked up and delivered to the warehouse.

Because of a variety in raw data comprising the performance measures, the data gathering approach at the second control point is somewhat more complex. The required information is:

- for activity 11: a description of the problem and the date the division notified the warehouse (complaint logbook);
- for activity 12: a description of what is done to resolve the issue and the date action is taken (complaint logbook); the time spent by a property specialist in resolving the specific issue versus the total time spent on all work activities during the issue resolution period (time card records); the total number of reports distributed during the measurement interval (property reports).

Step 7: *Analyze/report actual performance*
In this step, we will explore some of the statistical classical techniques to analyze and to display the results of the performance indicators to clearly communicate the answer to management questions. One way to look at the indicators is to use a bar chart to plot their progress over time, and to show some possible trends.

Fig. 5.44, 5.45, 5.46, 5.47, and 5.48 show the bar charts related to the 5 indicators defined at Step 4. Data refer to a five months monitoring period.

Step 8: *Compare actual performance to goal/standard*
In this step, we need to compare actual performances to the goals.

In this specific case, Fig. 5.44, 5.45, 5.46, 5.47, and 5.48, show that some objectives have not been met. Consequently, we should determine if the difference between each indicator and the target is not significant, or if corrective actions are necessary.

Step 9: *Definition of corrective actions*
In this step, we need to take the necessary action to bring the process performance back into line with goal(s). Typically, the key objectives of correction are:

- removal of defects and defect causes;

- attainment of a new state of process that will prevent defects from happening;
- maintenance or enhancement of the efficiency and effectiveness of the process.

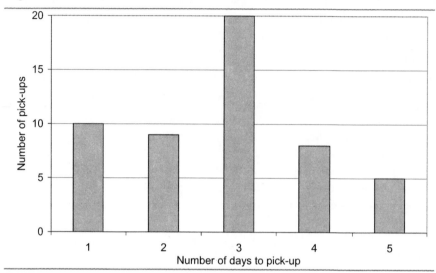

Fig. 5.44. Frequency chart related to indicator 1-A (*number of days elapsed from call to pick-up*), with reference to the 2nd week of the monitoring period

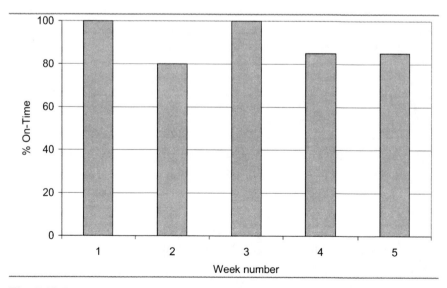

Fig. 5.45. Frequency chart related to indicator 1-B (*percentage of specially scheduled pick-ups on time*), with reference to the first 5 weeks of the monitoring period

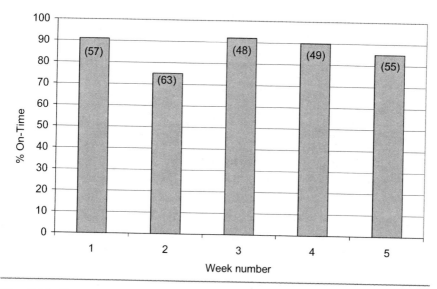

Fig. 5.46. Frequency chart related to indicator 1-C (*percent on-time pick-ups*), with reference to the first 5 weeks of the monitoring period (the total number of pick-ups in brackets)

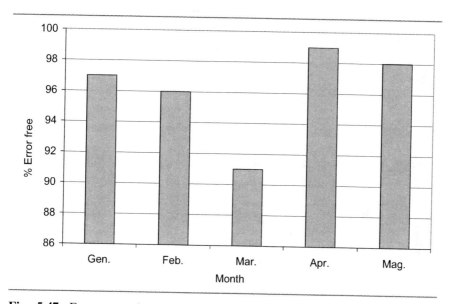

Fig. 5.47. Frequency chart related to indicator 2-A (*percent of accuracy of monthly report*), with reference to the first 5 weeks of the monitoring period

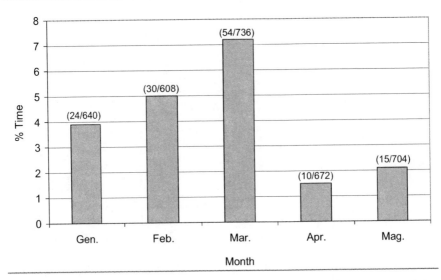

Fig. 5.48. Frequency chart related to indicator 2-B (*percent time spent resolving goods list problems*), with reference to the first 5 weeks of the monitoring period (the hours spent on resolutions and the total hours worked per month in brackets)

In this case study, for example, the goal of 95% on-time pick-ups is never met. As a consequence, it is necessary to find the root cause of the problem and to identify possible solution(s).

5.7 Maintaining a performance measurement system

Every performance measurement, in order to meet the organizational targets, need to be constantly maintained and – possibly – improved. With reference to the key components of a performance measurement system, the most critical ones are:

- the strategic plan;
- the key sub-processes;
- stakeholder needs;
- a possible change of the organizational framework;
- the occurrence of new regulations or standards;
- the possibility of make use of new support technologies;
- employee involvement.

If we find some changes within these components, it is necessary to perform a process alignment with organization's strategic objectives. For example, the emergence of a new competitor or a new program within the organization may impact the stakeholders' points of view or expectations. As a consequence, a part of the performance measurement system may need to be changed.

5.8   Effective use and misuse of indicators

This section provides examples of ways in which indicators, if misused, may be irrelevant or may encourage wrong activities resulting in less attention to outcome and quality. Furthermore, sometimes indicators – by themselves – are not appropriate for assessing outcomes, for determining future directions or for resource allocation. Pushing this concept to the limit, some authors state that indicators may not result in improved performance or a focus on outcomes. On the contrary, they may hamper the development and implementation of effective strategy. (Winston 1993; Mintzeberg 1994; Perrin 1998). According to Perrin (1998), the main causes of these problems and limitations are:

- *Varying interpretations of the "same" terms and concepts*
  Indicators, independently on what they represent, are invariably used, recorded and interpreted in varying ways. Thus, there can be a lack of comparability across different sites and staff, even with a seemingly straightforward measure such as the "number of client served". For example, what is a "client" in a retail outlet? Is it anyone who phones or walks in the door to receive some information about products? Or is it someone who has a regular relationship with the outlet? If the same individual goes to different sales points of the same chain of shops, is he counted as one or multiple clients? Many aggregated indicators can be interpreted in various ways, frequently inconsistent.

  Problems such as the above can be minimized by recognizing that virtually any indicator can be interpreted in various ways, by pre-testing indicators to identify how they may be interpreted in the field, and by providing a clear and an unambiguous definitions. Training and orientation also helps in leading to common interpretations, as does the active involvement and sense of responsibility of the staff. However, these techniques have limitations and are not always feasible. This is particularly the case when common measures are applied across a number of different agencies, or even different settings within the same organizations. Given staff turnover, orientation, training and monitoring process regarding compilation of indicators needs to be never ending to ensure quality control.

Another aspect to consider is that when staff feel that the future of their program or even their own jobs may be dependent upon "making the numbers look good", they inevitably will interpret definitions in a way that is most favourable to the agency. Perrin (1998) mentions an interesting example. Canada has employment equity legislation requiring federally regulated industries to develop and implement plans for equity in the employment of disadvantaged groups, including women, visible minorities, aboriginal people, and people with disabilities. One bank showed strong improvement in its record of employment of people with disabilities – until a legal centre representing the rights of people with disabilities grew suspicious and, upon investigation, discovered that the bank selectively changed its definition of disability to include a broader range of people, such as those requiring eyeglasses, as disabled, thus increasing their members.

- *Goal displacement*
When indicators become the objective, they result in "goal displacement", which leads to emphasis on the wrong activities, thus encourages means of "making the numbers" without improving actual outcomes. As a result, they frequently distort the direction of programs, diverting attention away from, rather than towards, what the program should be doing.

- *Use of meaningless and irrelevant measurements*
Indicators frequently do not reflect what is really occurring for many different reasons. Perrin (1998) describes an example about some court house clerks who pointed out how impossible it was to collect the information required for their reports. When Perrin asked how they compile and submit their weekly statistics, he was told that: "We put down something that sounds reasonable". Such situations should come as no surprise to anyone who has spent significant time in the field. But people such as some head office staff, senior managers, and even some evaluators removed from the realities of how things often work at the front lines often believe "the members", which have a seductive reality by their (false) precision.

Indicators can be irrelevant, even if they are accurate. The essence of a performance measurement system is to reduce a complex program to a small number of indicators (see Sect. 5.5). Indicators ignore the inherent complexities of social phenomena, which involve many interacting factors that cannot meaningfully be reduced to one or a limited number of quantitative indicators. In other terms, there is the problem of representing all the important dimensions of a process (property of exhaustiveness – Sect. 4.6.2). There is an inverse relationship between the importance of an indicator and the ease, or even possibility, of its quantification. As Patton (1997) has indicated: *"To require goals to be clear, specific and measurable is to require programs to attempt only those things that social scientists know how to measure"*. Many activities in the public policy realm, by their very nature, are complex and *intangible* and cannot be reduced to a numerical figure (Franceschini 2001). As Mintzberg (1996) stated: *"Assessment of many of the most common activities in government (or complex processes) requires soft judgement – something that hard measurement cannot provide ... Measurement often misses the point, sometimes causing awful distortions"*. The above discussion emphasizes that what is measured, or even measurable, often bears little resemblance to what is relevant.

* *Cost savings vs cost shifting*
    Indicators typically look at individual processes out of context, ignoring inter-relationships and the reality that few processes do – or should – act in isolation. Consequently, "outcomes" may represent cost shifting rather than true cost savings, ignoring or transferring needs and clients elsewhere rather than actually addressing them. For example, the number of "drop-outs" is a rather common indicator of the success of an Academic System. A low number of drop-outs indicate the Course effectiveness. Few systematic attempts, however, are made to discover why students leave, for how long and where they go. In some cases, individuals are merely transferred from an Academic Course to another. Students have not dropped out the Academic System. As a consequence, the number of drop-outs is not necessarily an indicator of lack of effectiveness.
    Indicators are invariably short term in nature. But short-term benefits and outcomes may result in future requirements and increased costs over the longer term, thus "shifting" costs into the future.

* *Misuse of derived indicators*
    Derived indicators may obscure subgroup differences. Consequently, the "same" outcome may reflect different forms of program effect. For example, the U.S. Census Bureau for 1996 showed an inflation-adjusted increase in income of 1.2% from the previous year. It indicates an improvement in the overall income of the people. However, a very different picture emerges if one examines the income of subgroups, for such an analysis reveals that the wealthiest 20% saw their income increase 2.2%, while for the poorest 20% it *decreased* 1.8%. In other words, the inequality gap, which has been widening for years, increased by a further 4% in a year when the economy was booming and unemployment falling. This finding would be totally missed by a derived indicator based upon aggregate family income.

- *Limitations of objective-based approaches to evaluation*
Three typical limitations of the use of objectives for evaluation purposes are the following:

  - it is difficult to distinguish between the ambitiousness and appropriateness of stated objectives on the one hand, and the actual results achieved on the other. Programs with ambitious objectives that "push the envelope" may be unfairly penalized, while mediocre programs attempting the commonplace are more likely to achieve their objectives;
  - objective-based approaches for evaluation do not take into account unintended or unanticipated effects and consequences, which may be both positive and negative and are frequently more important than the program's stated objectives;
  - objectives – and indicators – are typically fixed, while the environment, needs and program activities are constantly changing. A responsive program should be changing its objectives and targets. It should be reviewing whether its intended outcomes are still desirable, or need modification, or replacement. Objectives and indicators usually should become out of date.

- *Useless for decision making and resource allocation*
The most frequently mentioned rationale for performance measurement is to provide for more informed decision making and budgeting. But performance measurement, by itself, is useless for this purpose. As Newcomer (1997) states: "*Performance Measurement typically captures quantitative indicators that tell what is occurring with regard to program outputs and perhaps outcomes but, in itself, will not address the how and why questions*". For example, a program may fail to meet its performance targets because the program theory is wrong, in which case it should be replaced with something else. But it also may fail to do so for a variety of other reasons such as: inappropriate targets or measures which are not identifying other possible program outcomes; faulty management or implementation, under (or over) funding, faulty statistics, and so on. Use of indicators incorrectly assumes causality, inferring that the identified outcomes are a direct result of program activities. As evaluators know, causality can only be assessed through use of an appropriate evaluation design that aims at understanding the "whys" and mechanisms by which outcomes are achieved.

  Thus, indicators provide no direct implications for action, unless other means are used to explore the reasons for results and the potential for future impact. Indeed, it is dangerous to make decisions about the future of programs based upon indicators alone, as this is likely to lead to inappropriate action.

- *Less focus on outcome*
Perrin (1998) thinks that the sad, supreme irony is that performance measurement systems typically lead to less – rather than more – focus on outcome, innovation and improvement. A narrow focus on measurement is inconsistent with a focus on change and improvement that requires constant questioning about what else can be done or done better. An indicators misuse may leads to impaired performance, an emphasis on justifying and defending what was done, and a reluctance to admit that improvement is needed.

Despite the problems and limitations identified above, performance indicators are indispensable for many activities like process evaluation, resources allocation, and comparison of complex systems. However, it is important to remark that they should always be carefully analysed, selected and used.

5.9   Indicators as conceptual technologies

So far, we have dealt with the indicator properties, the construction of performance measurement systems, and their potential to represent a process. This section introduces another key to the reading of indicators role and impact. As already discussed (Sect. 1), indicators can influence the organizations which use them; in the following discussion we try to analyse this concept in more depth.

Indicators can be considered as *conceptual technologies*, able to intangibly influence organizations (Barnetson and Cutright 2000). Technology is seen as an ensemble of technical procedures and instruments to generate products or services. The adjective *conceptual* refers to the concept of intangibility.

On the basis of this assumption, indicators can shape *what* issue we think about and *how* we think about those issues through the selection and structure of the indicators that are used. This idea is derived from Polster and Newson's (1998) study concerning the external evaluation of the academic work. For example, a performance indicator that measures graduates' employment rates indicates to institutions that this outcome is of importance to the agency that mandated its introduction; the act of measurement makes institutional performance on this indicator public. By focusing institutional attention on their indicator performance, governments may impose a policy agenda on institutions by embedding assumptions related to purposes, goals or values into the selection and structure of indicators. In this way, performance indicators shift the power to set priorities and goals to those who create and control these documentary decision-making systems, thereby reconstructing the relationship between academics and those who construct and operate indicators systems.

In addition to *what*, indicators can also be used to shape *how* we think about an issue. The inclusion of indicators that demonstrate the positive outcomes of a policy agenda and the exclusion of indicators that demonstrate negative outcomes generates evidence that legitimize a particular policy agenda. Consequently, the use of indicators affects how institutions and policies are evaluated because the power to delineate what evidence is considered relevant is shifted to those who create and control indicator-driven systems. The use of indicators, however, not only shifts decision making power upwards and outwards, it also facilitates the use of financial rewards and punishments in order to manipulate institutional behaviour.

To examine in detail the overall impact of a system of indicators, Barnetson and Cutright (2000) suggested the conceptual model in Table 5.26. This analysis of the indicators impact is based on six dimensions: value, definition, goal, causality, comparability and normalcy.

Table 5.26. Six dimensions to evaluate indicators impact (Barnetson and Cutright 2000). With permission

Impact Dimension	Description
Value	The act of measurement delineates what activity or outcome is valued. That is, the inclusion or exclusion of indicators determines what is considered important and unimportant.
Definition	Performance indicators (re)define concepts (e.g., accessibility, affordability, quality, etc.) by operationalizing them in measurable terms.
Goal	Performance indicators include a point of reference by which a performance is judged. Performance indicators assign goals through both the value embedded in an indicator and the point of reference used in the indicator.
Causality	Performance indicators assign responsibility for an activity or outcome by embedding an assumption of causality.
Comparability	The use of common indicators assumes institutions (departments, individuals etc.) are comparable. This may pressure institutions to generate common outcomes or undertake common activities which may or may not be appropriate given institutional circumstances and mission.
Normalcy	Performance indicators delineate a range of normal behaviours or outcomes. This may pressure institutions to alter their activities so as to decrease a systemic disadvantage or increase a systemic advantage.

By making explicit the assumptions embedded in a series of indicators, it becomes possible to understand the broader policy agenda that underlies the indicators and, subsequently, could knowingly approve of, alter or critically challenge their implementation. Table 5.27 presents a series of questions designed to bring out the possible impact of each single indicator. Table 5.28 presents a series of questions designed to bring out the possible impact of a system of indicators.

Analysing the indicators impact may also illuminate unstated goals, how those goals are operationalized and attempts within the system to mitigate or alter goals at the operational level.

Table 5.27. Questions designed to bring out the possible impact of each single indicator (Barnetson and Cutright 2000). With permission

Impact Dimension	Questions
Value	By its inclusion, what does this indicator indicate is important to those who constructed and/or operate this indicator?
Definition	How does this indicator define a concept by operationalizing it in measurable terms? What alternative definition(s) of this concept exist?
Goal	What outcome does this indicator expect from an institution (department, individual, etc.) based upon the value and the point of reference embedded within it?
Causality	Who does this indicator make responsible for a performance? What assumption of causality underlies this assignment of responsibility? For example, making institutions responsible for the graduates' satisfaction assumes that institutions can control and deterministically influence the factors contributing to satisfaction.
Comparability	In what ways does this indicator assume institutions are comparable? For example, measuring external revenue generation by colleges, universities and technical institutes implies that rough parity in the ability of each type of institution to generate external revenue.
Normalcy	What assumptions does this indicator make about "normal" behaviours or outcomes? For example, measuring graduates' employment rate in fields related to their area of study at a fixed point after graduation assumes that it is desirable and possible for all graduates to find work within their disciplines and that graduates of all disciplines have roughly similar career trajectories.

It is generally accepted that performance indicators make knowledge "objective" – that is, independent of its creators and users through quantification (Porter 1995). *Quantified* knowledge is independent because it is less dependent than *narrative-derived* knowledge upon context for interpretation and, therefore, is more easily transported across time and distance with minimal loss of content. According to Porter, quantification constrains the ability of others to exercise judgment when they use the information thereby subordinating personal bias to public standards. Such *mechanical objectivity* (i.e., following a set of rules to eliminate bias) is similar to the political and moral use of objectivity to mean impartiality and fairness. This differs from *absolute objectivity* (i.e., knowing objects as they really are) and *disciplinary objectivity* (i.e., reaching consensus with one's peers about the nature of objects).

Table 5.28. Questions designed to bring out the possible impact of a system of indicators (Barnetson and Cutright 2000). With permission

Impact Dimension	Questions
Value	Do the system's indicators indicate what is important to those who construct and/or operate it?
Definition	Are there definitional trends evident within the system? For example, do the indicators in a system operationalize performances in economic terms?
Goal	Are there trends in the goals assigned by this system? For example, do the indicators consistently reward institutions that decrease government costs by increasing efficiency and broadening the funding base?
Causality	Is responsibility consistently attributed to one group? For example, a system of indicators may consistently assign responsibility for outcomes to institutions or it may disperse responsibility among several groups (e.g., government, students, institutions, exogenous environmental factors, etc.). Are there trends in the assumptions of causality that underlie the assignment of responsibility? For example, a system of indicators may assumes that institutions can control and deterministically influence the factors contributing to several indicators.
Comparability	How does the indicator system deal with comparisons between institutions? For example, a system of indicators may consistently (or inconsistently) recognize or ignore differences between institution's goals, missions, circumstances and resources.
Normalcy	What activities and/or outcomes does this system assume to be normal?

In this way, the use of indicators and performance funding in governance is designed to increase objectivity by applying a commonly agreed upon sets of rules to achieve a series of ends. In theory, then, the application of these measures should increase the impartiality of governance as decisions can be made based upon facts rather than other considerations. The belief that increasing objectivity (through quantification of outcomes) and linking resource allocation to outcomes will increase organizational effectiveness is consistent with the mechanical model of organizational functioning (Power 1996).

We can contest the objectivity of indicators by asserting that normative assumptions are embedded in indicators and shape *what* issues that we think about and *how* we think about them. This suggests that indicators are not a mere technical means of evaluating performance and/or allocating funding, but rather are a policy instrument designed to generate a particular set of outcomes (Barnetson and Cutright 2000).

As a policy instrument, indicators have the potential to significantly reduce institutional autonomy (i.e., the freedom to make substantive decisions about institutional direction). This view differs from the usual assertion that the use of indicators results in much needed institutional accountability. Using indicators to shape institutional behaviour (particularly when indicators guide resource allocation) confuses accountability with *regulation* (Kells 1992). *Regulation* involves an outsider examining a performance and acting to maintain or change it (possibly through rewards and/or punishments). Regulation erodes autonomy rather than promoting it.

To study the consequences of the introduction of a performance measurement system, we consider the example of the Academic funding system in Alberta (Canada), which is based upon nine performance indicators (Barnetson and Cutright 2000). Five indicators are used by all institutions (the *learning component*) while four indicators affect only research universities (the *research component*). Each institution's total score is used to allocate funding (AECD 1999).

The learning component's five indicators fall into three categories based upon the government's goals of increasing *responsiveness, accessibility* and *affordability* (AECD 1997, 1999). Institutional responsiveness to the needs of learners and to provincial social, economic and cultural needs are assessed by examining the *employment rates* of graduates and graduates' *satisfaction* with their educational experience. Institutional progress towards higher levels of accessibility (i.e., increasing the number of students enrolled) is indicated by examining changes in full-load equivalent (FLE) enrolment based on a three-year rolling average. Institutions' success at maintaining affordability (i.e., providing quality learning opportunities to the greatest number of Albertans at a reasonable cost to the learner and taxpayer) is indicated by examining *administrative expenditures* and *outside revenue* generated.

Table 5.29 shows the indicators definition, while Table 5.30 shows their possible impact (only for *learning component*).

As the example above has shown, evaluating the impact of indicators is complex. The scheme of Barnetson and Cutright (2000) is an interesting contribute towards this direction. We can note that, while physical systems are regulated by natural laws independent of the model considered, organized system are influenced by the way in which they are analysed and modelled (Hauser and Katz 1998).

Table 5.29. Indicators selected for evaluating the Academic funding system in Alberta (Canada). Indicators concern the learning component only (AECD 1997). With permission

Employment rate: percentage of graduate-survey respondents employed within a specified period following program completion

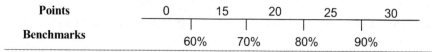

Points	0	15	20	25	30
Benchmarks		60%	70%	80%	90%

Graduate satisfaction with overall quality: percentage of graduate-survey respondents fully/somewhat satisfied with overall educational quality

Points	0	15	20	25	30
Benchmarks		70%	80%	90%	95%

Credit FLE: percentage change in full-load equivalent enrolment from one period to the next

Points	0	20	25	30
Benchmarks	Urban	-2%	0%	+4%
	Rural	-5%	0%	+4%

Administrative expenditures: administration as a percentage of total expenditures less ancillary expenditures

Points	0	3	4	5
Benchmarks	>3500 students	11%	7%	5%
	<3500 students	12%	8%	6%

Enterprise revenue: revenues less all government grants, tuition fees under policy, sponsored research (universities only), ancillary services and earned capital contributions as a percentage of government grants

Points	1	3	4	5
Benchmarks	Urban	20%	35%	50%
	Rural	10%	25%	40%

Table 5.30. Analysis of the impact of the indicators selected for evaluating the Academic funding system in Alberta (Canada) (Barnetson and Cutright 2000). Third column concerns the specific assumptions embedded in each indicator. With permission

Indicator	Impact Dimension	Specific description
Employment rate	Value	High levels of graduate employment are desirable.
	Definition	Responsiveness entails matching programming to labour market needs.
	Goal	Institutions should increase graduates' employment rates.
	Causality	Institutions can (1) control program offerings and (2) match program offerings with labour market demands.
	Comparability	Institutions are equally able to generate labour market outcomes.
	Normalcy	All institutions' graduates' have comparable career trajectories.
Graduate satisfaction	Value	High levels of graduate satisfaction are desirable.
	Definition	Responsive entails providing programs that satisfy graduates.
	Goal	Institutions should increase the satisfaction rate of their graduates.
	Causality	Institutions can control the factors that contribute to graduates' satisfaction.
	Comparability	Institutions are equally capable of satisfying their learners.
	Normalcy	An institution's graduates have compatible program expectations.
Credit FLE	Value	Enrolment growth is desirable.
	Definition	Accessibility is a function of student spaces (measured by enrolment).
	Goal	Institutions should increase their enrolment.
	Causality	Institutions can influence (1) the demand for spaces and (2) the availability of spaces.
	Comparability	Institutions are equally able to increase enrolment.
	Normalcy	Economies of scale are equal between institutions.
Administrative expenditures	Value	Low levels of administrative expenditures are desirable.
	Definition	Affordable entails minimizing administrative expenditures.
	Goal	Institutions should decrease administrative expenditures.
	Causality	Institutions can control the factors that contribute to administrative expenditures.
	Comparability	Institutions face similar economies (and diseconomies) of scale.
	Normalcy	Enrolment increases reduce per-student administrative costs.

Table 5.30. (cont.)

Enterprise revenue	Value	High levels of non-government/non-tuition revenue are desirable.
	Definition	Affordability entails maximizing external revenue generation.
	Goal	Institutions should increase external revenue generation.
	Causality	Institutions can generate external revenue.
	Comparability	Institutions have similar abilities to generate external revenue.
	Normalcy	Raising revenue is compatible with institutions' teaching function.

6. Indicators, measurements, preferences and evaluations: a scheme of classification according to the representational theory

6.1 Introduction[9]

The properties of an object, either directly or indirectly observable, may essentially be judged and described by three basic operations: *measurements, evaluations* and *preferences*. The aim of the present chapter is to propose a scheme to classify these three operations, and to provide a comparison within the indicators environment.

The question has a particular scientific importance and some possible repercussions involve many disciplines like metrology, decision-making, and quality measurement. Considering this latter aspect, when we ask a sample of people to express opinions about the quality of a good, we carry out an operation which is hovering among measurement, preference and evaluation. It is necessary to know what the conceptual paradigms at heart of these three operations are. The *representational theory of measurement* based on the properties of binary relations is the instrument used for this investigation.

Given a set of objects or alternatives A, a binary relation R on A is a subset of the Cartesian product $A \times A$. Binary relations arise very frequently from everyday language: for example, if A is the set of all people in a certain country, then the set:

$$T = \{(a, b): a, b \in A \text{ and } a \text{ is brother of } b)\} \tag{6.1}$$

defines a binary relation on A, which we may call "brother of".

The properties (reflexivity, symmetry, transitivity, etc...) of a generic relation R must be defined on a specific reference set. For example, the re-

---

[9] Special thanks to Dr. Paolo Cecconi for the draft of the present chapter (Cecconi et al. 2006)

lation "brother of" is symmetric on the set of all males in a certain country, but it is not symmetric on the set of all people in that country.

Formally, rather than a simple relation, we speak of a *relational system* (*A, R*), that is a relation applied to a set of objects (see Chap. 3). This concept, introduced by Tarski (1954), has been the natural vehicle for the subsequent development of the representational theory of measurement (Scott and Suppes 1958; Krantz et al. 1971, 1989, 1990).

When a person judges or describes objects (physical or abstract), he/she considers *one* or *more comparison relations*. These relations may be tangible or intangible, uniformly or not uniformly interpretable. For example, the relations "*more beautiful than*", "*more elegant than*", "*worthier than*" "*preferred to*" are relations intangible and arbitrarily interpretable by different subjects, whereas the relations "*heavier than*", "*longer than*", "*warmer than*" are not. These latter are observable relations, which do not enable a free interpretation by the observers. There is a direct reference to the scales of the International System of measurement.

In the following description, the analysis of the differences among three operations takes place considering only a specific property of the objects. "*The subjects of measurement are properties. Of course, properties exist only in connection with empirical objects. Usually, one object shows various properties. In measuring one property, we neglect all the other properties the object in question may have*" (Pfanzagl 1968).

Moreover, the considerations here referred to are exclusively limited to rankings. We only consider the cases in which the individuals are able to establish a priority ranking, that is a hierarchy among objects.

6.2   Two criteria of discrimination: empiricity and objectivity

The Representational Theory notions, partially introduced in Chap. 3, will be extended to the concepts of *evaluation* and *preference*.

"*Measurement is the assignment of numbers to properties of objects or events in the real world by means of an objective empirical operation, in such a way as to describe them. The modern form of measurement theory is representational: numbers assigned to objects/events must represent the perceived relations between the properties of those objects/events*" (Finkelstein and Leaning 1984).

The definition of a measurement refers to two fundamental concepts: empiricity and objectivity.

The **empiricity** arises when a judgment of a ranking "*is the result of observations and not, for example, of a thought experiment. Further, the concept of the property measured must be based on empirically determinable relations and not, say, on convention*" (Finkelstein 2003).

There is empiricity when the type of relation is observable, that is the property of the object proves to be, in a precise moment, in a well defined state characterized without ambiguity. Empiricity means that there is "*an objective rule for classifying some aspect of observable objects as manifestations of the property*" (Finkelstein 1982).

The **objectivity** concerns the kind of results that the judgment produces, "*within the limits of error independent of the observer*" (Finkelstein 1982). "*Experiments may be repeated by different observers and each will get the same result*" (Sydenham et al. 1989). Full objectivity means independence of the subjects. The result of the operation gives only information about the measured property.

The **measurement** requires both empiricity and objectivity. It is an operation of objective description: the results of n different measurements, in the same operating conditions, are univocal and independent by subjects. We suppose there is no "error" and uncertainty in an ideal measurement process (the environmental and other influential variables are considered nonexistent). It is also an empirical operation: "*Measurement has something to do with assigning numbers that correspond to or represent or "preserve" certain observed relations*" (Roberts 1979).

The **preference** is neither empirical nor objective. Preferences are, by definition, subjective and conflicting. We are not able to know exogenously, in detail, the relation that each subject applies when assigning a ranking. An outside observer will have considerable difficulties in interpreting the results generated by this kind of operation. In other words, different subjects interpret the relation in different ways and can establish disagreeing orderings. In this case, the uncertainty concerns deeply the kind of relation applied by each individual.

The **evaluation** is somewhere between measurement and preference. It is not objective because evaluations are individual perceptions, performed without the use of an univocal instrument of measurement. Nevertheless, it is an operation that *wants* to be empirical: the meaning of intangible relations is circumscribed by means of an exogenous process of semantic definition from the outset. Subjects are called on to conform to this process. Operatively, there is uncertainty in the interpretation that subjects give to the provided dimension of observation.

The three operations can be classified as illustrated in the Table 6.1:

Table 6.1. Scheme of classification of the three operations

	Objective	Empirical
Measurement	Yes	Yes
Evaluation	No	Yes
Preference	No	No

Before continuing, it is convenient to give anexplicit explanation of the following terms:

- *exogenous*: the expression of a description/ordering is subordinated to a coactive, explicit and declared constraint. An outside observer imposes rules (concerning the dimension of observation, interpretation of the scales, etc…) to which subjects conform from the outset.
- *endogenous*: the expression of a description/ordering happens according to a latent, implicit, non-declared point of view. Each subject decides to adopt the rules he considers more convenient, without declaring them.

6.3 The representational definition of measurement

The assumption that the relations are observable is at the heart of a representational point of view.

In a measurement, subjects are called on to "judge" an observable relation, on which there are no doubts about meaning and interpretation. Some possible examples of these relations are: "*longer than*", "*heavier than*", "*warmer than*", etc...

As explained in Chap. 3, the representational theory of measurement related to a quality or a property of an object has four fundamental parts:

- **An empirical relational system.** Consider some quality (for example the length of an object) and let a_i represent an individual manifestation of the quality A, so that we can define a set of all possible manifestations as A={a_1, …}. Let there be a family of empirical relations R_i on A, R={R_1,…,R_n}. Then the quality can be represented by an empirical relational system $\mathfrak{A} = \langle A, R \rangle$.
- **A symbolic/numerical relational system.** Let Z represent a class of numbers Z={z_1, …}. Let there be a family of relations P={P_1, …P_n} defined on Z. Then $\mathfrak{Z} = \langle Z, P \rangle$ represents a numerical relational system.
- **A representation condition.** Measurement is defined as an objective empirical operation such that $\mathfrak{A} = \langle A, R \rangle$ is mapped omomorphically into (onto) $\mathfrak{Z} = \langle Z, P \rangle$ by M and F. Specifically, *M* is the function map-

ping A to Z, so that $z_m = M(a_m)$ (M: A → Z). F is the function mapping one-to-one the relations of R on P (F: R → P).

The above **homomorphism** is the representation condition. Firstly it implies that if a_n is related to a_m by an empirical relation R_k, that is $R_k(a_n, a_m)$, P_k is the numerical relation corresponding to R_k, $z_m = M(a_m)$ is the image of a_m in Z under M then $R_k(a_n, a_m)$ implies and is implied by $P_k(z_n, z_m)$.

Measurement is a *homomorphism* - not an *isomorphism* - because M is not a one-to-one function. It maps separate but indistinguishable property manifestations into the same number. The definition of a representational measurement is illustrated in Fig. 6.1.

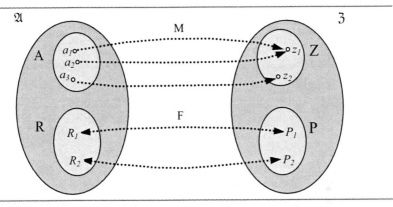

Fig. 6.1. Homomorphism of 𝔄 onto 𝔍. Two elements in A may be mapped into the same number in Z (Roberts 1979)

> "In measurement we start with an observed or empirical relational system and we seek a mapping to a numerical relational system which "preserves" all the relations and operations observed in the empirical one" (Roberts 1979).
>
> "Whatever inferences can be made in the numerical relational system apply to the empirical one" (Dawes and Smith 1985).

- **Uniqueness condition.** The representation condition may be valid for more than one mapping function M. There are admissible transformations from one scale to another scale without invalidating the representation condition. The uniqueness condition defines the class of transformations for which the representation condition is valid (Finkelstein and Leaning 1984; Franceschini 2001; Finkelstein 2003). For ordinal scales, all monotone increasing functions are admissible transformations.

As discussed in Sect. 3.2, indicators do not fulfil the condition of uniqueness. In fact, compared to measurements, they are defined under less restrictive hypothesis.

6.3.1 An example of ordinal measurement

A measurement can be viewed as a representation of some properties of the real world by symbols; nominal and ordinal scales are examples of representation by non numeric symbols (Franceschini 2001; Finkelstein 2003).

The ability to order a set of objects/events is related to the notion of the "amount" of a property manifested by each one. In ordinal measurement, the relation "$\succ$" ("*it has got the property more than*") is transitive and reflexive. Therefore:

$$q \succ r \Leftrightarrow M(q) > M(r) \tag{6.2}$$

being q and r two manifestations of the examined system.

For example, let us consider the hardness measurement of minerals. The relation R "*harder than*" is a typical example of ordinal measurement. In this case the ranking is based on a symbolic assignment, rather than a numeric one (see Sect. 3.2.2).

The Mohs scale orders minerals from diamond to talc, on the basis of which scratches which. The ability to scratch (i.e. to etch, to cut in surface) is the empirical relation and the ordering is the formal relation.

The scale is built as follows: ten standard minerals are arranged in an ordered sequence so that precedent ones in the sequence can be scratched by succeeding ones and cannot scratch them. The standards are assigned numbers 1 to 10 (symbols).

The sequence is *talc*-1, *gypsum*-2, *calcite*-3, *fluorite*-4, *apatite*-5, *orthoclase*-6, *quartz*-7, *topaz*-8, *corundum*-9, *and diamond*-10.

A mineral sample of unknown hardness which cannot be scratched by quartz and cannot scratch it, is assigned measure 7 (Finkelstein 1982).

The *homomorphism* is a faithful representation of empirical relations by symbolic relations. The condition of homomorphism – namely of an assignment of symbols (typically numbers) to objects/events according to the degree of presence of a certain property – is considered by the followers of the representational theory as a necessary and sufficient condition to define a measurement.

Some authors (Dawes and Smith 1985) assert that not all possible rules of assignment yield right measurements. The assignment of numbers is a representational measurement only if the three following requirements are satisfied:

- be orderly;
- represent meaningful attributes;
- yield meaningful predictions.

According to the authors, the presence of these three conditions defines the *automatic consistency check*, which realizes the difference between a representational measurement and nonrepresentational one: *"when mineral (a) scratches mineral (b), then (a) is represented above (b) in the order"* (Dawes and Smith 1985).

If consistency check fails, we cannot speak about representational measurements. Consider, for example, a subject expressing his/her own opinion about a given product. The judgment is expressed in a *rating* scale of the type indicated in Fig. 6.2. In this particular case, suppose that the subject selects the label "+3".

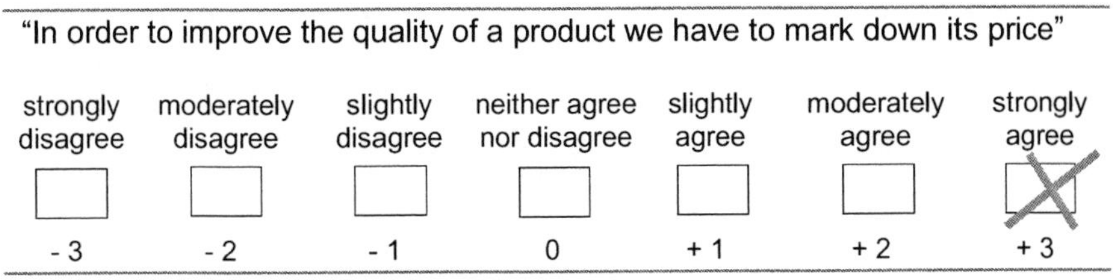

Fig. 6.2. An example of a measurement with no consistency

This statement *"doesn't represent a specific behaviour"* of the subject. In fact, we may wish to make many inferences on the basis of this behaviour. For example that he/she will turn his/her own attention to those firms adopting this policy, that he/she is deceiving himself/herself about something that will not happen, or that he/she believes in paying suppliers by manufacturing money. *"But we cannot make a firm prediction about some other response to this or another rating scale. There is no consistency check, hence no representational measurement"* (Dawes and Smith 1985).

Vice versa, measurements of hardness, mass, length, etc. present a consistency check. This check is performed by an appropriate measurement system realizing the homomorphism from the empirical relational system to the numerical (symbolic) one.

The measurement system guarantees the two fundamental components of a measurement: "assignment and empirical determination" (Mari 1997). It circumscribes without ambiguity the empirical relations (*harder than, heavier than*, *longer than*, etc...), and performs the assignment of numbers to the objects according to the rule of the corresponding homomorphism.

The presence of a conventional, non ambiguous, empirical reference is the reason at the base of the objectivity of measurement results.

6.4 Evaluations

In general, the qualities of an object/event can be classified as *physical* or *non physical*, *observable* or *indirectly observable*. Trying to measure non physical and non directly observable magnitudes, many operational problems arise. In these situations – with a few exceptions – we cannot speak about *real* measurement, but just about evaluations or attributions of values to individual judgments. Due to the lack of common reference standards, descriptions about non tangible qualities are possible only by means of subjective judgments, expressed on adequate scales.

Fig. 6.3 illustrates the concept of evaluation of a *non physical* magnitude. The subject ideally compares the object (first scale pan) with the reference terms on the scale (second scale pan). The evaluation consists in identifying the judgment on the scale which balances the two scale pans.

This way, Pawson (1997) asserts that: "*First, evaluation deals with the real, that is we evaluate things and empirical relations about things. Secondly, evaluation should follow a realist methodology. Thirdly, evaluation, perhaps above all, needs to be realistic*".

6.4.1 Psychophysical evaluations

Physical magnitudes measurements fulfil the properties of *empiricity* and *objectivity*. However, if they are performed through the perceptions of single individuals, rather than using a proper measure, they may not fulfil objectivity.

Let A be a set of objects and R the relation "*heavier than*". That is, for any pair of object x, y in A we define:

$$x \, R \, y \Leftrightarrow x \text{ is heavier than } y \quad (6.3)$$

"*Note that R can be defined either by a balance or by psychophysical experiments using an observer to compare the weights. The two procedures yield similar empirical relational systems with the same object set. The interpretation of the relation "heavier than", however, is physical in the former system and psychological in the latter*" (Coombs et al. 1970).

In the second case, we speak of evaluations, or – more precisely – psychophysical evaluations.

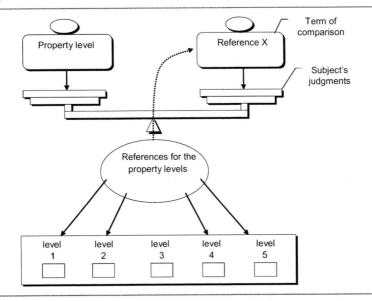

Fig. 6.3. Basilar scheme of evaluation of a non physical magnitude (Franceschini 2001)

6.4.2 The evaluation of non tangible qualities

In general we speak about evaluations when we attempt to measure non tangible attributes like utility, attitude, or a non tangible performance property.

The description of these attributes or latent constructs is not objective. It may originate in a free interpretation of meaning by the subjects. In these cases, the first fundamental component of a measurement (empiricity) runs short.

The first problem is *"the difficulty of establishing an adequate objective concept, or theoretical construct, of these qualities based on empirical operations"* (Finkelstein 1982).

Consider, for example, the aesthetical beauty of an object: in this case *"there is not an objective rule for classifying some aspect of observable objects as manifestations of the beauty. Similarly, there are no objective empirical relations such as indistinguishability or precedence, in respect* of beauty. The basis for the measurement of beauty is thus absent from the outset" (Finkelstein 1982).

The aim of an evaluation is that of "building" and "imposing" to evaluators, in some way, this empiricity component.

According to the presented classification, most of the *scaling* techniques (used in the modern psychophysics, from Stevens – 1951 – onwards) should be considered as evaluations to "measure" non directly observable properties (Torgerson 1958). A scale is constituted by a group of *items*, to discriminate objects or events, from the viewpoint of the examined property.

The evaluation is typically a normative process. It is based on exogenous rules driving subjects in the attribution of values to intangible and interpretable qualities. Essentially, there are two kinds of rules:

- clear and precise definition of non physical attributes (abstract constructs);
- definition of operating evaluation scales.

The first fundamental step of an evaluation process consists in the definition of a *reference axis*. The evaluation process needs to make the initial latent construct observable and less interpretable, giving to it "*empirical substance*", by specific empirical rules.

The second phase consists in providing suitable evaluation scales. Depending on the type of evaluation, many different scale types can be used (Franceschini 2001).

An evaluation requires that there is "uniformity" for subjects in accepting the rules provided. This uniformity does not exist in preference judgments, where everyone is free to interpret the situation in their own manner.

6.4.3 Evaluation: a subjective homomorphism

Psychological tests or Questionnaires for the evaluation of product or service quality are examples of evaluation processes. In these cases we speak about evaluations rather than measurements, because the objectivity may not be fulfilled. Paraphrasing measurement definition, the evaluation becomes the assignment of numbers, or labels to properties or events of the real world by means of an empirical subjective operation, in such a way as to describe them.

We said that evaluation is a partially empirical and subjective operation. We are going to justify this position by means of the representational theory.

In the evaluation context, the mapped relation R is not empirical (as it is for measurements). The relation becomes empirical by means of a set of semantic (what the construct means) and operating rules (evaluation scale). We exogenously impose the dimension or relation that subjects have to observe.

What is perceived as "*x has this property more than y*" in the empirical world, has an immediate translation in the ranking performed by the subject. In ordinal evaluation, there is a homomorphism from the empirical world onto the symbolic (numeric) one, that is to say, there is a faithful (even if subjective) representation of empirical objects and their relations in the numerical world. The representational form is, therefore, maintained. From this point of view, the homomorphism is <u>not</u> able to formalize the difference between the operation of evaluation and measurement (Mari 2003).

The question is that we deal with a subjective operation. The ranking performed by a subject will not coincide with that of another subject. It is not univocal. Different subjects can observe different degrees of properties in the same object. This dependence may cause some important formal consequences.

The concepts of indifference threshold δ and the distinction between the indifference relation I (reflexive and symmetric) and the equivalence relation E (reflexive, symmetric and transitive) are strictly connected with each subjective operation. This latter distinction supports the difference between evaluation and measurement. There is a relation of indifference between two stimuli-objects (for example, two sounds of different intensity) every time each subject is not able to discriminate amongst them. The equivalence relation E is stronger than the indifference relation I. It does not confine itself simply to asserting the non difference between two elements (for example the two auditory stimuli mentioned before), but it establishes their equality (Roberts 1979).

The different significance of the two relations is fundamental in psychological/psychophysical evaluations. Imagine three weights (a, b and c); if, weighing them by hand, $a\,I\,b$ (a is indifferent to b) and $b\,I\,c$, it cannot be said that also $a\,I\,c$. A little difference between a and b and between b and c can become noticeable when the conjoint effects of the single differences are considered. The absence of transitive property in the indifference relation shows this possibility. This property, on the other hand, belongs to the equivalence relation E: if $a\,E\,b$ (a is equivalent to b) and $b\,E\,c$, then $a\,E\,c$.

The problem of discrimination between the equivalence and indifference relation is connected to the concept of "*Just Noticeable Differences*", originally introduced by Fechner (1860), and to the definition of *semiorders* (Luce 1956).

Just Noticeable Differences arise when different stimuli are judged indifferent from a subject who is not able to discriminate among them. These differences become noticeable using measurement instruments more accurate than a simple subjective evaluation. The homomorphism performed by the single subject will not coincide with the homomorphism performed by means of an adequate measurement system.

6.4.4 Problems and questions still open

Some fundamental problems arise from ordinal evaluations:

* *Dimension of representation "compatible" with the way of thinking of subjects*

The priority aim of an evaluation consists in yielding *empirical predictions*, for example, about the purchasing intentions of customers, the undertaking of an investment, etc. Therefore, it is necessary that the provided *dimension* of representation is compatible with the way of thinking of the subjects, otherwise the evaluations will be neither significant nor useful.

Setting up a good evaluation scale means having to know the dimension of representation which is most important for the subjects. *"Clearly, not all rating scales are compatible with intuitive thought, nor does compatibility imply that rating scales are isomorphic with such thought"* (Dawes and Smith 1985).

* *The design of a scale reflecting the real capacity of discrimination of subjects*

This problem mostly arises from *rating* scales with enumerated categories, where the subject expresses himself on a verbal category and not on a linear *continuum*. In fact, problems of interpreting the meaning of categories can subsist. A verbal label, *"very satisfying"*, may not have the same meaning for both a very demanding subject and one who is easy to please. This poses a real problem to codify information, as the scale of interpretation adopted by each subject is usually unknown (Franceschini 2001).

The response category can be interpreted by subjects as too wide or too narrow. When the category is too wide, the subject is forced to classify as equal objects those he perceives as slightly different. In this case, there are no representational measurements, because a faithful representation of observed relations in numerical relations does not occur: *"it is essential that the relations among the objects of the world be properly reflected by the relations among the numbers assigned to them"* (Coombs et al. 1970).

The opposite problem can also occur. The response categories could be too detailed and the subject is confused in giving an answer. He may find himself in more than one response category. In this case the single subject's evaluations are not significant in absolute terms, but they are in relative (ordering) terms (see the Example 4.15).

The adequate definition of the width and of the meaning of each scale category brings evaluations to the stage of *homomorphisms* from an empirical relational system to a numerical one, that is to say, *representational evaluations*. This result remains an actionable target only after *"numerous interactions with evaluator subjects"* during the setting up phase of the scale (Finkelstein 1982).

- *The aggregation of evaluations expressed by many subjects*

This problem is connected with any subjective operation. The aggregation of individual rankings in a *"social"* global ranking is object of study of many disciplines: social and behavioural sciences (*Social Choice Theory*), operational research and economics (Arrow 1963; Fishburn 1973; Keeney and Raiffa 1976; Roy 1996).

6.5 Preference

Preference is the act by which, in presence of two or more possible objects, one of them can be chosen over the other, because it is considered more pleasant, more convenient and more conform to ones own tastes, interests, ideals, etc.

"Preference is necessarily relative to a subject. A preference is always somebody's preference. A preference, moreover, is relative not only to a subject but also to a certain moment or occasion or situation in the life of a subject. Not only may have different people with different preferences, but one and the same man may revise his preferences in the course of his life...the concept of preference is related to the notion of betterness" (Wright 1963).

"Preferences, to a greater or lesser extent, govern decisions...into our axiomatic system an individual's preference relation on a set of alternatives enters as a primitive or a basic notion. This means that we shall not attempt to define preferences in terms of other concepts...preferences between decision alternatives might be characterized in terms of several factors relating to the alternatives" (Fishburn 1970).

When a subject says that he/she prefers alternative a to b, he/she makes a relation between a and b which seems perfectly mouldable with the mathematic notion of a binary relation.

Suppose A is a collection of alternatives among which you are choosing, and suppose

$$P=\{(a,b)\in A\times A: \text{you (strictly) prefer } a \text{ to } b\}; \qquad (6.4)$$

Then P is called (strict) preference relation.

If A is a set of alternatives, $a\,P\,b$ holds if and only if you prefer (strictly) a to b, it is possible to assign a real number $u(a)$ to each $a\in A$, such that for all $a, b\in A$,

$$aPb \Leftrightarrow u(a) \succ u(b) \qquad (6.5)$$

The function u is often called ordinal utility function.

This assignment allows the relation "*preferred to*" of the single subject to be observed.

Nevertheless, as Roberts asserts, "*often, "preferred to" doesn't define a relation*" (Roberts 1979, pp 272-273).

With these remarks, Roberts wants to underline the absolute peculiarity of this relation that enjoys neither the property of consistency nor that of transitivity.

Consider the following example: a subject is called on to vote among three candidates A, B and C. If the subject prefers alternative A to alternative B and B to C, he/she will not necessarily prefer alternative A to C, as the transitivity property requires. It can occur that C is preferred to A (see Sect. 3.3.1).

The subject assigning preferences could have an implicit model of preferences such that it can not be mapped on any numerical structure (it can contain intransitivity chains). Without transitivity, it is not possible to establish an ordering.

The lack of transitivity often arises for intangible relations like "*preferred to*", "*more beautiful than*", "*more elegant than*", etc... relations arbitrarily interpretable, because they are not directly observable.

6.5.1 The impossibility of the representational form (for preferences)

What differences emerge when comparing the definition of preference with that of measurement?

In general, we are faced with a preference assignment every time the subject is called on to perform a ranking amongst things without a "measurement system" or a set of predefined rules, such as in the evaluation process. The preference ranking among alternatives is the result of an endogenous activity of a subject who chooses, in an arbitrary way, how to represent the relation. Preference becomes *"an arbitrary measurement conceived as a decision-making activity"* (Sartori 1985).

Examples of arbitrarily interpretable relations are: *"worthier than"*, *"better than"*, *"more beautiful than"*, etc...

The assignment of preferences is neither empirical, nor objective. It does not deal with an empirical operation because the subject chooses arbitrarily the relation considered remarkable for the ranking. The above choice is completely endogenous, different from subject to subject. We are not able to understand exogenously what is the dimension of representation selected and followed by the subject and the interpretation he/she has given to it.

There is not a transfer of observable relations into numerical relations. We cannot speak of a homomorphism as defined by a representational point of view. There is no mapping onto a numerical relational system *"preserving both relations (and operations) observed in the empirical relational system"* (Roberts 1979).

Paraphrasing a Stevens' expression, preference can be defined as *"a measurement according to any rule"* (Stevens 1951). Subjects may have chosen to observe one of the infinite possible relations on the objects in order to perform a preference ranking. It is obvious that this operation is completely subjective: *"the measurement value is not so much a property of the thing measured as something which expresses an appreciation of the measurer towards the thing itself. What counts, does not count because it counts in itself, but because it is judged to count by someone"* (Mari 1997)".

When subjects report their own opinions about constructs which are not adequately detailed, an attribution of values to individual preference judgments takes place.

Constructs like the utility of a service, the aesthetic of a product, the guidability of a vehicle, must be specified and detailed to transform preferences into evaluations. Otherwise, each subject will interpret the construct as he/she considers more convenient. The reason for this is a *"semantic ambiguity"* in the constructs (Dawes and Smith 1985). In these cases, they speak about *"nonrepresentational measurements"*. As an example, consider the responses to the following two attitude questions:

- *"The adoption of a Certified Quality System is a necessary burden"*.
- *"Advantages of a Certified Quality System are larger than disadvantages"*.

A positive answer to both *items* seems to indicate a favourable attitude of the management toward quality certification, but how should that response be interpreted and represented? Who answers "yes" has a mildly positive attitude. Those with either a strongly positive or strongly negative attitude would answer "no". In contrast, an affirmative response to the latter *item* should be interpreted as meaning that the responders' favourability surpasses neutrality, but how far we do not know. *"The point is that they involve choice based on semantic knowledge, our semantic knowledge. There is nothing in the observation of affirmative answers themselves that dictates how they are to be interpreted and represented"* (Dawes and Smith 1985).

The presence of this ambiguity is the main difference between preference and evaluation. In an evaluation process we will try to circumscribe this semantic ambiguity, fully defining in this case the concepts of "necessary burden", "advantage" and "disadvantage".

We say that a preference has not the empiricity and objectivity requirements of a measurement. However, the Representational Theory considers that representational measurements of preference are possible. It is important to concisely define this position. When subjects assign a ranking to the elements of a certain set, they make their own preferences explicit and consequently their own relation *"preferred to"* on that set is made observable. According to the representational theory, this assignment is a real representational measurement. It is viewed as a homomorphic representation of the relation *"preferred to"* from the empirical world (even if the relation is not explicit in the empirical domain).

Roberts (1979) presents the case of preferences among classical music composers. The author points out that it is possible to speak of representational measurements of individual preferences, when these satisfy the axioms of Cantor's theorem. Given a set of objects and a defined relation of preference (P), the relational system (A, P) has to satisfy the following conditions:

asymmetry: if $aPb \Rightarrow \neg bPa, \forall a,b \in A$

(6.6)

(natural property of preferences)

negative transitivity: if $\neg aPb$ & $\neg bPc \Rightarrow \neg aPc, \forall a,b,c \in A$ (6.7)

Axioms of Cantor's theorem are necessary and sufficient conditions to have ordinal representational measurements.

However, enacting Roberts' position, *"some information can be obtained on the evaluating subject, about their way of seeing things, but surely not on the elements of the set, i.e. the empirical world"* (Mari 2003).

Roberts does not dwell on the analysis of the meaning of the relation but he/she simply analyses and interprets the results of the representation. The author seems to neglect that the relation *"preferred to"* is not empirical because it is interpretable and therefore observable in an arbitrary way.

The problem is that we do not know what is the meaning of the relation *"preferred to"*, in terms of empirical relations. Therefore, an empirical relational system for preference cannot be identified.

6.6   The concept of "dictation"

The classification in Table 6.1 can be represented through the scheme in Fig. 6.4.

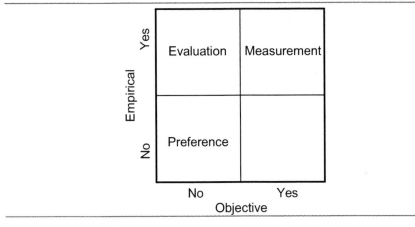

Fig. 6.4. Classification scheme of classification of the three operations: measurements, evaluations and preferences

Referring to Fig. 6.4, a box still remains empty. This is the occurrence of "Yes" for "Objective" and "Not" for "Empirical". We can define this situation as a new operation called *dictation*.

We have a dictation when the mapping between the empirical system and the symbolic system is defined in order to give a predefined result, independently from the occurrence of manifestations in the empirical space. This operation reflects the intervention of a *dictator* who aprioristically "dictates" the result.

A dictation is objective because it always produces the same result, independently from the subject who performs the mapping. It is not empirical because there is not a set of empiric rules to produce the result (see Fig. 6.5).

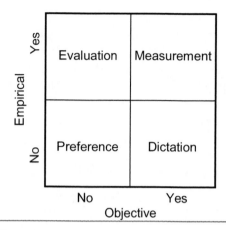

Fig. 6.5. Classification scheme of measurements, evaluations, preferences and dictations

This is the case, for example, in fixed competitions, where the winner is aprioristically defined and the selection rules are arbitrarily proposed in order to obtain the imposed result.

Furthermore, dictations can be interpreted as a particular case of indicators. They also represent a map from an empirical system onto a symbolic system (see Fig. 6.6).

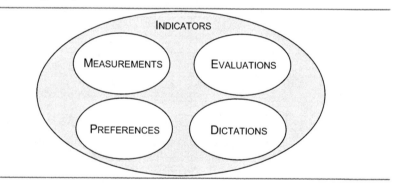

Fig. 6.6. Measurements, evaluations, preferences and dictations can be considered as a subset of indicators

6.7 Conclusions

In general, objects and events are described and classified on the basis of specific indicators. Measurements, evaluations, preferences and dictations are the four basic operations which can be used for this purpose. The difference between them lies in the way they treat relations among objects/events.

In the case of preference, relations are thought by subjects and transferred directly to the result of the representation under an endogenous decision-making activity, different from subject to subject. The inability to exogenously know the way in which things have been interpreted by subjects leads to a total absence of empiricity and objectivity in the operation.

Measurements set the empirical relations that must be represented without ambiguity by means of a measurement system which realizes the *"internal consistency check"* and makes feasible the fundamental requirement of objectivity.

On the other hand, it is the aim of the evaluation process to provide all possible "tools" in order to reduce the problems of semantic interpretation of the *items* and of the scale categories. In this way, individual judgments are treated as representational evaluations.

The notion of homomorphism is effective in marking the difference between preference and measurement, but it does not define a clear border between the operation of measurement and evaluation. The source of the difference between the two operations is that evaluations from the outset allow for the possibility of choosing different dimensions of representation.

Representational theory enables to give a precise and effective definition of empiricity in the three operations, but not of objectivity. Some authors have emphasized the inability to incorporate the fundamental requirement of objectivity of a measurement in the formalization of the approach as the limit of a representational point of view. Due to this, some claim the necessity of an operational-representational approach for the sake of an exhaustive definition of measurement, with the introduction of a measurement system (Mari 2003).

According to the proposed scheme of classification, a further operation, the *dictation*, can be defined. Dictation is objective but not empirical: it produces a predefined (by a dictator) result, independent of the subject who performs the mapping.

Particular attention should be posed to the problem of aggregation of individual evaluations or preferences into "social" or group results. Problems of aggregation do not appear in the field of measurements because of their objective nature. On the contrary, an abundant literature about the aggregation of preferences is available. In this framework and renowned for its relevance, remains the *Arrow's Impossibility Theorem* (1963) which represents the milestone of *Social Choice Theory*. By means of the theory of relations, Arrow proved the non-existence of aggregation mechanisms for individual preferences in a social preference, where four fundamental axioms are satisfied: *Unrestrained Domain of Preferences, Independence of irrelevant alternatives, Pareto weakness, and Non-dictatorship* (Arrow 1963).

With the aim of tracing a possible path for a future research, the traditional problem of aggregation of individual preferences can be reformulated in the problem of aggregation of individual evaluations. This would allow us to see if the imposition of exogenous bonds relaxes Arrow's conditions. In substance, it is natural to wonder if Arrow's Theorem is still valid when the input is changed from preferences into "canalized" preferences as evaluations. Arrow's theorem can be adopted as a second possible instrument of investigation to further highlight differences between preferences and evaluations.

References

AECD (1997), Advanced Education and Career Development, Rewarding Progress Towards Goals: Highlights of Alberta's Performance Envelope Funding., Edmonton (Canada).

AECD (1999), Advanced Education and Career Development, 1998/99 Annual Report, Edmonton (Canada).

AKAO V. (1988), Quality Function Deployment, Productivity Press, Cambridge (MA).

ARPA - Piedmont's Regional Agency for the Environment (2005), "Tabella dei valori massimi in Torino - gennaio 2005", ARPA environmental vigilance and control network, Turin, Italy.

ARROW K. (1963), Social Choice and Individual Values, 2nd edn, New Haven, Yale University Press.

ATKINSON A. (1997), Linking Performance Measurement to Strategy, «Journal of Strategic Performance Measurement», vol 1, n 4, pp 5.

ATKINSONS A.B. (1970), On the measurement of inequality, «Journal of Economic Theory», vol 2, pp 244-263.

AUSTIN R.B. (1996), Measuring and Managing Performance in Organizations, Dorset House Publishing, New York.

AZZONE G. (1994), Innovare il sistema di controllo di gestione, Etas Libri, Milano, Italy.

BALDRIGE NATIONAL QUALITY PROGRAM (BNQP) (2006), Criteria for performance Excellence, NIST Publications, Gaithersburg (MD), site: http://www.baldrige.nist.gov.

BANKER R. D., CHANG H., JANAKIRAMAN S. N., KONSTANTS C. (2004), A balanced scorecard analysis of performance metrics, «European Journal of Operations Research», vol 154, pp 423-436.

BARBARINO F. (2001), UNI EN ISO 9001:2000. Qualità, sistema di gestione per la qualità e certificazione, Il Sole 24 Ore, Milan.

BARNETSON B., CUTRIGHT M., (2000), Performance indicators as conceptual technologies, «Higher Education», vol 40, pp 277-292.

BEAUMON B.M. (1999), Measuring supply chain performance, «International Journal of Operations and Production Management», vol 19, n 3, pp 275-292.

BELLANDI G. (1996), La misurazione della qualità, Etas Libri, Milan.

BORDA J. C. (1781), Mémoire sur les élections au scrutin, Comptes Rendus de l'Académie des Sciences. Translated by Alfred de Grazia as Mathematical derivation of an election system, Isis, vol 44, pp 42-51.

BOUYSSOU D., MARCHANT T., PIRLOT M., PERNY P., TSOUKIÀS A., VINCKE P. (2000), Evaluation and decision Models: A Critical Perspectives, Kluwer Academic Publishers, Boston, MA.

BROWN M. G. (1996), Keeping Score: Using the Right Metrics to Drive World-Class Performance, Quality Resources, New York, NY.

BROWN M. G. (1999), Baldrige Award Winning Quality: How to Interpret the Malcolm Baldrige Award Criteria. Ninth Edition, ASQ Quality Press, Milwaukee.

CAPLICE C., SHEFFI Y. (1994), A Review and Evaluation of Logistics Metrics, «The International Journal of Logistics Management», vol 5, n 2, pp 11-28.

CAPLICE C., SHEFFI Y. (1995), A Review and Evaluation of Logistics Performance Measurement Systems, «The International Journal of Logistics Management», vol 6, n 1, pp 61-64.

CECCONI P., FRANCESCHINI F., GALETTO M. (2006), Measurements, evaluations and preferences: A scheme of classification according to the representational theory. «Measurement», vol 39, n 1, pp 1-11.

CIMOSA (1993), Open System Architecture for CIM, Edition Esprit, Consortium AMICE Editor, Springer Verlag, Berlin, site: www.cimosa.de.

COMUNITÀ EUROPEA (2002), Direttiva 2002/3/EC del Parlamento Europeo e del Consiglio del 12 febbraio 2002 relativa all'ozono nell'aria, Brussels.

CONDORCET M. J. A. N. C. (Marquis de) (1785), Essai sur l'application de l'analyse à la probabilité des décisions redues à la pluralité des voix, Imprimerie Royale, Paris.

COOMBS C.H., DAWES R.M., TVERSKY A. (1970), Mathematical Psychology: an Elementary Introduction, Prentice-Hall, Englewood Cliffs, New Jersey.

DAWES R. M., SMITH T.L (1985), Attitude and Opinion Measurement. In: Lindzey G., Aronson E. (ed.), Handbook of Social Psychology, vol 1, Random House, New York.

DENTON D. K. (2005), Measuring relevant things, «International Journal of Productivity and Performance Management», vol 54, n 4, pp 278-287.

DISPEA (2004-05), DIPARTIMENTO DI SISTEMI DI PRODUZIONE ED ECONOMIA DELL'AZIENDA, Project: "Il nuovo Call Center del Comune di Torino: monitoraggio della qualità della comunicazione offerta ai cittadini" (in alliance with AICQ and Comune di Torino).

DISPEA (2005), DIPARTIMENTO DI SISTEMI DI PRODUZIONE ED ECONOMIA DELL'AZIENDA, Project: "Valutazione della qualità nei servizi SO (Service Operations)" (in alliance with RAI).

DIXON J.R., NANNI Jr., ALFRED J., VOLLMANN T.E. (1990), The New Performance Challenge: Measuring Operations for World-Class Competition, Dow Jones-Irwin, Homewood, IL.

DRAFT FEDERAL INFORMATION (1993), Announcing the Standard for "Integration Definition For Function Modelling (IDEF0)", Processing Standards Publication 183, site: www.idef.com.

EDWARDS A. L. (1957), Techniques of Attitude Scale Construction, Appleton-Century, New York.

EDWARDS J. B., (1986), The Use of Performance Measures, Montvale, NJ.

EUROPEAN FOUNDATION FOR QUALITY MANAGEMENT (2006), site: www.efqm.org.

EISENHARDT K.M. (1989), Agency Theory: An Assessment and Review, «The Academy of Management Review», vol 14, n 1, pp 57-74.

ERNST and YOUNG (1990), International Quality Study: The Definitive Study of the Best International Quality Management Practices, American Quality Foundation, New York.

EU COMMISSION (1996), Directive 96/62/EC of the European Parliament, Official Journal of the European Communities.

EVANS J. R. (2004), An Exploratory Study of Performance Measurement Systems and Relationship with Performance Results, «Journal of Operations Management», vol 22, pp 219-232.

FECHNER G. (1860), Elemente der Psychophisik, Breitkopf und Hartel, Leipzig.

FINKELSTEIN L. (1982), Handbook of measurement science, Theoretical Fundamentals, Sydenham P.H. (ed), vol 1, John Wiley & Sons, New York.

FINKELSTEIN L. (2003), Widely, strongly and weakly defined measurement, «Measurement», vol 34, n 1, pp 39-48.

FINKELSTEIN L., LEANING M. (1984), A Rewiew of the fundamental concepts of Measurement, Gonella (ed.), Proc. of the Workshop on fundamental logical concepts of Measurement, IMEKO, 1983, reprinted in: Measurement, vol 2, n 1.

FISHBURN P. (1970), Utility Theory for Decision Making, John Wiley & Sons, New York.

FISHBURN P. (1973), The Theory of Social Choice, Princeton University Press, New Jersey.

FISHBURN P. C. (1976), Noncompensatory preferences, «Synthese», vol 33, pp 393-403.

FISHBURN P. C. (1977), Condorcet social choice functions, «SIAM Journal on Applied Mathematics», vol 33, pp 469-489.

FISHBURN P. C. (1982), The foundations of expected utility, D. Reidel, Dordrecht.

FLAPPER S.D.P., FORTUIN L., STOOP P.P.M. (1996), Toward consistent performance measurement systems, «International Journal of Operations & Production Management», vol 16 , n 7, pp 27-37.

FRANCESCHINI F. (2001), Dai Prodotti ai Servizi. Le nuove frontiere per la misura della qualità, UTET Libreria, Torino.

FRANCESCHINI F. (2002), Advanced Quality Function Deployment, St. Lucie Press/CRC Press LLC, Boca Raton, FL.

FRANCESCHINI F., GALETTO M. (2004), An empirical investigation of learning curve composition laws for Quality Improvement in complex manufacturing plants, «Journal of Manufacturing and Technology Management», vol 15, n 7, pp 687-699.

FRANCESCHINI F., ROMANO D. (1999), Control chart for linguistic variables: a method based on the use of linguistic quantifiers, «International Journal of Production Research», vol 37, n 16, pp 3791-3801.

FRANCESCHINI F., GALETTO M., MAISANO D. (2005), A short survey on Air Quality Indicators: Properties, Use, and (Mis)use, «Management of Environmental Quality: An International Journal», vol 16, n 5, pp 490-504.

FRANCESCHINI F., GALETTO M., VARETTO M. (2005), Ordered Samples Control Charts for Ordinal Variables, «Quality and Reliability Engineering International», vol 21, n 2, pp 177-195.

FRANCESCHINI F., GALETTO M., MAISANO D. (2006), Classification of Performance and Quality Indicators in Manufacturing, «International Journal of Services and Operations Management», vol 2, n 3, pp 294-311.

FRANCESCHINI F., GALETTO M., MAISANO D., VITICCHIÈ L. (2006), The Condition of Uniqueness in Manufacturing Process Representation by Performance/Quality Indicators, «Quality and Reliability Engineering International», vol 22, n 5, pp 567-580.

GALBRAITH J.R. (1973), Designing Complex Organizations, Addison-Wesley, Reading, MA.

GALBRAITH L., GREENE T. J. (1995), Manufacturing System Performance Sensitivity to Selection of Product Design Metrics, «Journal of Manufacturing Systems», vol 14, n 2, pp 71-79.

GALBRAITH L., MILLER W. A., GREENE T. J. (1991), Pull System Performance Measures: A Review of Approaches for System Design and Control, «Production Planning and Control», vol 2, n 1, pp 24-35.

HALLIDAY D., RESNICK R. (1996), Fundamental of Physics - Part 1, 5th ed., John Wiley & Sons, New York.

HAUSER J., KATZ G. (1998), Metrics: You Are What You Measure!, «European Management Journal» vol 16, n 5, pp 517-528.

HENDRICKS K.B., SINGHAL V. R. (1997), Does Implementing an Effective TQM Program Actually Improve Operating Performance? Empirical Evidence from Firms That Have Won Quality Awards, «Management Science», vol 43, n 9, pp 1258-1274.

HILL T. (1999), Manufacturing Strategy: Text and Cases, McGraw-Hill, Burr Ridge, IL.

HOEK, G., BRUNEKREEF, B., GOLDBOHM, S., FISCHER, P., VAN DEN BRANDT, P. A. (2003), Cardiopulmonary mortality and air pollution, «The Lancet», vol 362, n 9384, p 587, Maryland, USA.

HOUGHTON COLLEGE (2004), Dashboard Indicators: The Top Mission-Critical Indicators for Houghton College, site: http://campus.houghton.edu/

IAAF - International Association of Athletics Federations (2005), Scoring Tables, site: www.iaaf.org.

ISO 9000:2000, Quality Management Systems - Fundamentals and Vocabulary, ISO, Geneva.

ITTNER C.D., LARCKER D.F. (1998), Innovations in performance measurement: trends and research implications, «Journal of Management Accounting Research», vol 10, pp 205-238.

JURAN J. M. (1988), Juran on Planning for Quality, The Free Press, New York.

JURAN J. M. (2005), Quality Control Handbook, 5th edn, McGraw-Hill, New York.

KAPLAN R. S., NORTON D. (1992), The Balanced Scorecard-Measures That Drive Performance, «Harvard Business Review», January-February, pp 71-79.

KAPLAN R. S., NORTON D. (1996), The Balanced Scorecard, Harvard Business School Press, Cambridge, MA.

KAPLAN R. S., NORTON D. P. (2001), The Strategy-Focused Organization: How Balanced Scorecard Companies Thrive in the New Business Environment, Harvard Business School Press, Boston, MA.

KAYDOS W. (1999), Operational Performance Measurement: Increasing Total Productivity, St. Lucie Press, Boca Raton, FL.

KEARNEY A. T. (1991), Measuring and Improving Productivity in the Logistics Process: Achieving Customer Satisfaction Breakthroughs, Council of Logistics Management, Chicago.

KEENEY R., RAIFFA H. (1976), Decisions with Multiple Objectives: Preference and Value Tradeoffs, Wiley & Sons, New York.

KELLS H. (1992), Self-regulation in Higher Education: A Multinational Perspective on Collaborative Systems of Quality Assurance and Control, Jessica-Kingsley, London.

KERR S. (2003), The best-laid incentive plans, «Harvard Business Review», January, pp 27-37.

KRANTZ D., LUCE R., SUPPES P., TVERSKY A., (1971, 1989, 1990), Foundations of Measurement, vol 1, (1971), vol 2. (1989), vol 3. (1990), Academic Press, New York.

LAMBERT D.M., BURDUROGLU R. (2000), Measuring and selling the value of logistics, «International Journal of Logistics Research», vol 11, n 1, pp 1-17.

LEONG G.K., WARD P.T. (1995), The Six Ps of Manufacturing Strategy, «International Journal of Operations and Production Management», vol 15, n 12, pp 32-45.

LING R.C., GODDARD W.E. (1988), Orchestrating Success: Improve Control of the Business with Sales and Operations Planning, Wiley, New York.

LINS M. P. E., GOMES E. G., SOARES DE MELLO J. C. B., SOARES DE MELLO A. J. R. (2003), Olympic Ranking Based on a Zero Sum Gains DEA Model, «European Journal of Operational Research», vol 148, n 2, pp 85-95.

LOCKAMY III A., SPENCER M.S. (1998), Performance measurement in a theory of constraints environment, «International Journal of Production Research», vol 36, n 8, pp 2045-20060.

LOHMAN C., FORTUIN L., WOUTERS M. (2004), Designing a performance measurement system: A case study, «European Journal of Operational Research», vol 156, pp 267-286.

LUCE R.D. (1956), Semiorders and a Theory of Utility Discrimination, «Econometrica», vol 24, pp 178-191.

LYNCH R.L., CROSS K.F. (1995), Measure Up! How to Measure Corporate Performance, Blackwell, Malden MA.

MAGRETTA J. STONE N. (2002), What Management is: How it Works and Why it's Everyone's Business, Free Press, New York.

MARI L. (1997), The role of determination and assignment in measurement, «Measurement», vol 21, n 3, pp 79-90.

MARI L. (2003), Epistemology of Measurements, «Measurement», vol 34, n 1, pp 17-30.

MASKELL B. H. (1991), Performance Measurement for World Class Manufacturing, Productivity Press, Cambridge, MA.

MAYER R.J., MENZEL C.P., PAINTER M.K., DE WITTE P.S., BLINN T., PERAKATH B. (1995), Information Integration for Concurrent Engineering (IICE): IDEF3 Process Description Capture Method Report, Human Resources Directorate Logistics Research Division, Texas A&M University.

MCKENNA R. (1997), Real Time: Preparing for the Age of the Never Satisfied Customer, Harvard Business School Press, Boston, MA.

MELNYK S. A., STEWART D. M., SWINK M. (2004) Metrics and performance measurement in operations management: dealing with the metrics maze, «Journal of Operations Management», vol 22, pp 209-217.

MELNYK S.A., CHRISTENSEN R.T. (2000), Back to Basics: Your Guide to Manufacturing Excellence, St. Lucie Press, Boca Raton, FL.

MENTZER J. T., KONRAD B. P. (1988), An efficiency/Effectiveness Approach to Logistics Performance Analysis, «Journal of Business Logistics», vol 12, n 1, pp 36-61.

MINISTÈRE DE L'ÉCOLOGIE ET DU DÉVELOPPEMENT DURABLE (2004), Arrêté du 22 julliet 2004 relatif aux indices de la qualité de l'air, «Journal official de la Règublique Française», France.

MINISTERO DELL'AMBIENTE E DELLA TUTELA DEL TERRITORIO (2002), Decreto 2 aprile 2002 n. 60, Recepimento della direttiva 1999/30/CE del Consiglio del 22 aprile 1999 concernente i valori limite di qualità dell'aria ambiente per il biossido di zolfo, il biossido di azoto, gli ossidi di azoto, le particelle e il piombo e della direttiva 2000/69/CE relativa ai valori limite di qualità dell'aria ambiente per il benzene ed il monossido di carbonio, Gazzetta Ufficiale n 87 del 13 Aprile 2002, Roma.

MINTZBERG H. (1994), The rise and fall of strategic planning, The Free Press, New York.

MINTZBERG H. (1996), Managing Government, Governing Management, «Harvard Business Review», May-June, pp 75-83.

MOCK T. J., GROVE H. D. (1979), Measurement, Accounting and Organizational Information, John Wiley & Sons Inc, NY.

NARAYANA C. L. (1977), Graphic positioning scale: an economical instrument for surveys, «Journal of Marketing Research», vol XIV, pp 118-122.

NATIONAL PARTNERSHIP FOR REINVENTING GOVERNMENT (1997), Serving the American Public: Best Practices in Performance Measurement. Federal Benchmarking Consortium Report. Washington, DC.

NEELY A. (1998) The Performance Measurement Revolution: Why Now and What Next?, «International Journal of Operations and Production Management», vol 18, pp 9-10.

NEELY A., GREGORY M., PLATTS K. (1995), Performance Measurement System Design, «International Journal of Operations and Production Management», vol 4, pp 80-116.

NEELY A., RICHARDS H., MILLS J., PLATTS K., BOURNEAL M. (1997), Designing performance measures: a structured approach. «International Journal of Operations & Production Management», vol 17, n.11, pp 1131-1152.

NEMHAUSER G.L., WOLSEY L.A. (1988), Integer and Combinatorial Optimization, John Wiley, New York.

NEVEM WORKGROUP (1989), Performance Indicators in Logistics, IFS Ltd. and Springer-Verlag, Bedford (UK).

NEW C.C., SZWEJCZEWSKI M. (1995) Performance measurement and the focused factory: empirical evidence, «International Journal of Operations and Production Management», vol 15, n 4, pp 63-79.

NEWCOMER K. (1997), (Ed.) Using performance measurement to improve public and nonprofit programs. New Directions for Evaluation, n. 75, Jossey-Bass, San Francisco.

NIST (National Institute of Standard and Technology) (2006), sites: http://www.quality.nist.gov and http://www.baldrige.nist.gov.

NUNALLY J. C., BERNSTEIN I.H. (1994), Psychometric Theory, 3rd edn, McGraw-Hill, NY.

NURMI H. (1987), Comparing voting systems, D. Reidel, Dordrecht.

OFFICE OF THE AUDITOR GENERAL OF CANADA AND THE COMMISSIONER OF THE ENVIRONMENT AND SUSTAINABLE DEVELOPMENT (1993), Developing Performance Measures for Sustainable Development Strategies.

PARKER R.G., RARDIN R.L. (1988), Discrete Optimization, Academic Press, San Diego.

PATTON M.Q. (1997), Utilization-focused evaluation: The new century text. Sage Publications Inc., Thousand Oaks, CA.

PAWSON R. (1997), Realistic Evaluation, Sage Publications, London.

PERFORMANCE-BASED MANAGEMENT SPECIAL INTEREST GROUP (PBM SIG) (2001), The Performance-Based Management Handbook, vol 1, Establishing and Maintaining A Performance-Based Management Program, Oak Ridge Institute for Science and Education, (ORISE) – U.S. Department of Energy.

PERFORMANCE-BASED MANAGEMENT SPECIAL INTEREST GROUP (PBM SIG) (2001), The Performance-Based Management Handbook, vol 2, Establishing an Integrated Performance Measurement System, Oak Ridge Institute for Science and Education, (ORISE), U.S. Department of Energy, site: http://www.orau.gov/pbm/pbmhandbook/pbmhandbook.html.

PERFORMANCE-BASED MANAGEMENT SPECIAL INTEREST GROUP (PBM SIG) (2001), The Performance-Based Management Handbook, vol 3, Establishing Accountability for Performance, Oak Ridge Institute for Science and Education, (ORISE) – U.S. Department of Energy.

PERFORMANCE-BASED MANAGEMENT SPECIAL INTEREST GROUP (PBM SIG) (2001), The Performance-Based Management Handbook, vol 4, Collecting Data To Assess Performance, Oak Ridge Institute for Science and Education, (ORISE) – U.S. Department of Energy.

PERFORMANCE-BASED MANAGEMENT SPECIAL INTEREST GROUP (PBM SIG) (2001), The Performance-Based Management Handbook, vol 5, Analyzing

And Reviewing Performance Data, Oak Ridge Institute for Science and Education, (ORISE) – U.S. Department of Energy.

PERFORMANCE-BASED MANAGEMENT SPECIAL INTEREST GROUP (PBM SIG) (2001), The Performance-Based Management Handbook, Vol 6, Using Performance Information To Drive Improvement, Oak Ridge Institute for Science and Education, (ORISE) – U.S. Department of Energy.

PERFORMANCE-BASED MANAGEMENT SPECIAL INTEREST GROUP (1995), How To Measure Performance – A Handbook of Techniques and Tools – U.S. Department of Energy, site:
http://www.orau.gov/pbm/documents/documents.html.

PERRIN B. (1998), Effective Use and Misuse of Performance Measurement, «American Journal of Evaluation», vol 19, n 3, pp 367-379.

PFANZAGL J. (1968), Theory of Measurement, John Wiley, New York.

PFEFFER J., SALANCIK G.R. (1978), The External Control of Organizations: A Resource Dependence Perspective, Harper & Row, New York.

PIEDMONT REGIONAL LAW 43/2000 (2000), "Disposizioni per la tutela dell'ambiente in materia di inquinamento atmosferico. Prima attuazione del Piano regionale per il risanamento e la tutela della qualità dell'aria", Piedmont Regional Council, Turin, Italy, site: www.regione.piemonte.it.

POLSTER C., NEWSON J. (1998), Don't count your blessing: The social accomplishments of performance indicators', in Currie J., and Newson J. (Editors), Universities and Globalization: Critical Perspective. Thousand Oaks: Sage, pp 173-182.

PORTER T.M., (1995), Trust in Numbers: The Pursuit of Objectivity in Sciences and Public Life. Princeton University Press, Princeton.

POWER M. (1996), Making things auditable, «Accounting, Organizations and Society», vol 21, n 2-3, pp 289-315.

PREMIO QUALITÀ ITALIA (2002), Guida alla Partecipazione e all'Autovalutazione, Associazione Premio Qualità Italia, Ministero delle Attività Produttive, Roma.

RAPPORT, D. J., HOWARD J., LANNIGAN R., McCAULEY W. (2003), Linking health and ecology in the medical curriculum, «Environment International», vol 29, n 2-3, pp 353-358.

RITZ, B., YU, F., FRUIN, S., CHAPA, G., SHAW, G. M., HARRIS, J. A. (2002), Ambient Air Pollution and Risk of Birth Defects in Southern California, «American Journal of Epidemiology», vol 155, n 1, pp 17-25.

ROBERTS F. S. (1979), Measurement Theory, Addison-Wesley Publishing Company, Reading Mass, New York.

ROY B. (1993), Decision science or decision-aid science, «European Journal of Operational Research», vol 66, pp 184-204.

ROY B. (1996), Multicriteria Methodology for Decision Aiding, Kluwer Academic Publishers, Dordrecht.

ROY B., BOUYSSOU D. (1993), Aide Multicritère à la Décision: Méthodes et Cas, Economica, Paris.

SARTORI S. (1991), Il procedimento di misurazione, «Ingegneria», n 5-6.

SCHMENNER R.W., VOLLMANN T.E. (1994), Performance measures: gaps, false alarms and the usual suspects, International Journal of Operations & Production Management», vol 14, n.12, pp 58-69.

SCOTT D., SUPPES P. (1958), Foundational aspects of theories of measurement, «Journal of Symbolic Logic», n 23, pp 113-118.

SKINNER W. (1974) The decline, fall, and renewal of manufacturing plants, «Harvard Business Review», May-June.

SMITH D. (2000), The Measurement Nightmare: How the Theory of Constraints Can Resolve Conflicting Strategies, Policies, and Measures, St. Lucie Press, Boca Raton, FL.

STEVENS S.S. (1951), Mathematics, Measurement and Psychophysics, S.S. Stevens (ed.), Handbook of Experimental psychology, pp 1-49, John Wiley, New York.

SUNYER J., BASAGANA X., BELMONTE J., ANTO J. M. (2002), Effect of Nitrogen Dioxide and Ozone on the Risk of Dying in Patients with Severe Asthma, «Thorax - International Journal of Respiratory Medicine», vol 57, n 8, pp 687-693.

SYDENHAM P.H., HANCOCK N.H., THORN R. (1989), Introduction to Measurement Science and Engineering, John Wiley & Sons, New York.

TARSKI A. (1954), Contributions to the Theory of Models, Indagationes Mathematicae, n 16 , pp 572-588.

THOMSON J., VARLEY S. (1997), Developing a Balanced Scorecard at AT&T, «Journal of Strategic Performance Measurement», vol 1, n 4, p 14.

THORBURN W. M. (1918), The Myth of Occam's razor, Mind, vol 27, pp 345-353.

TORGERSON W.S. (1958), Theory and Methods of Scaling, John Wiley, New York.

ULRICH K., EPPINGER S. (2000), Product Design and Development, McGraw-Hill, New York.

U. S. DEPARTMENT OF ENERGY (1996), Guidelines for Performance Measurement (DOE G 120.1-5), site: http://www.orau.gov.

U. S. DEPARTMENT OF ENERGY (1996), Office of Environmental Management (DOE/EM), Office of Environmental Management Critical Few Performance Measures-Definitions and Methodology.

U. S. GENERAL ACCOUNTING OFFICE (1998), Performance Measurement and Evaluation: Definitions and Relationships, GGD-98-26.

U. S. OFFICE OF MANAGEMENT AND BUDGET (1995), Primer on Performance Measurement, site: http://govinfo.library.unt.edu/npr/library/omb/22a6.html.

U.S. DEPARTMENT OF ENERGY/NEVADA OPERATIONS – DOE/NV (1994), Performance Measurement Process Guidance Document, DOE/NV Family Quality Forum, June.

U.S. DEPARTMENT OF THE TREASURY (1994), Criteria for Developing Performance Measurement Systems in the Public Sector.

UNI 11097 (2003), Indicatori e quadri di gestione della qualità, Milano.

UNITED NATIONS DEVELOPMENT PROGRAMME (1997), Human Development Report 1997, Oxford University Press, Oxford.

UNITED NATIONS DEVELOPMENT PROGRAMME (2003), Human Development Report 2003, Oxford University Press, Oxford.

UNITED STATES ENVIRONMENTAL PROTECTION AGENCY (1999), Guideline for Reporting of Daily Air Quality – Air Quality Index, EPA-454/R-99-010 Press, North Carolina.

UNIVERSITY OF CALIFORNIA (1999), Laboratory Administration Office, Appendix F, Objective Standards of Performance, site: http://labs.ucop.edu/internet/lib/lib.html.

UNIVERSITY OF CALIFORNIA (1999), Laboratory Administration Office, Seven Years of Performance-Based Management: The University of California/Department of Energy Experience, site: http://labs.ucop.edu/internet/lib/lib.html.

VANSNICK, J.C (1986), De Borda et Condorcet à l'aggregation Multicritere, «Ricerca Operativa», vol 40, pp 7-44.

VICARIO G., LEVI R. (1998), Calcolo delle probabilità e statistica per ingegneri, Esculapio, Bologna.

VINCKE P. (1992), Multi-criteria decision aid, John Wiley, New York.

WINSTON J. (1993), Performance Indicators: Do they perform?, «Evaluation News and Comment», vol 2, n 3, pp 22-29.

WINSTON J. A. (1999), Performance Indicators – Promises Unmet: A Response to Perrin, «American Journal of Evaluation», vol 20, n 1, pp 95-99.

WRIGHT G. (1963), The logic of preference, University Press, Edinburgh.

YANG, I. A., HOLZ, O., JÖRRES, R. A., MAGNUSSEN, H., BARTON, S. J., RODRÍGUEZ, S., CAKEBREAD, J. A., HOLLOWAY, J. W., HOLGATE, S. T. (2005), Association of Tumor Necrosis Factor - Polymorphisms and Ozone-induced Change in Lung Function, «American Journal of Respiratory and Critical Care Medicine», vol 171, pp 171-176.

ZARNOWSKY F. (1989), The decathlon – A colourful history of track and field's most challenging event, Leisure Press, Champaign.

Index